Bio-Inspired Manufacturing Systems

Scrivener Publishing
100 Cummings Center, Suite 541J
Beverly, MA 01915-6106

Advances in Production Engineering

Series Editor: Dr. Sandip Kunar (sandip.sandip.kunar@gmail.com)

"Advances in Production Engineering" addresses recent developments and research issues related to the production engineering which necessitates the development of new materials and manufacturing processes in order to produce high-quality products at lower prices and in less time. This series aims to provide a scientific platform for researchers, practitioners, professionals, and academicians to discuss the most recent technological developments in metals, polymers, ceramics, composites, biomaterials, nanomaterials, special materials, metals, micro-forming, powder metallurgy, ceramics processing, non-traditional machining, high speed machining, micro and nanomachining, and laser processing. Tribological analysis, friction behavior, modelling, and optimization techniques in materials, machining, and manufacturing are also covered in the series.

Publishers at Scrivener
Martin Scrivener (martin@scrivenerpublishing.com)
Phillip Carmical (pcarmical@scrivenerpublishing.com)

Bio-Inspired Manufacturing Systems

Jagadeesha T.
Dept. of Mechanical Engineering, National Institute of Technology, Calicut, Kerala, India

and

Sandip Kunar
Dept. of Mechanical Engineering, Aditya University, Surampalem, Andhra Pradesh, India

Scrivener Publishing

WILEY

This edition first published 2026 by John Wiley & Sons, Inc., 111 River Street, Hoboken, NJ 07030, USA and Scrivener Publishing LLC, 100 Cummings Center, Suite 541J, Beverly, MA 01915, USA

For more information about Scrivener publications please visit www.scrivenerpublishing.com.

Wiley Global Headquarters
111 River Street, Hoboken, NJ 07030, USA

For details of our global editorial offices, customer services, and more information about Wiley products visit us at www.wiley.com.

The manufacturer's authorized representative according to the EU General Product Safety Regulation is Wiley-VCH GmbH, Boschstr. 12, 69469 Weinheim, Germany, e-mail: Product_Safety@wiley.com.

AI disclosure statement: The author acknowledges the use of AI-based tools for minor language editing or grammatical correction. However, all intellectual contributions, analyses, and conclusions presented in this chapter are entirely the author's own.

Library of Congress Cataloging-in-Publication Data

ISBN 978-1-394-40833-7

Cover image: Generated with AI using Adobe Firefly
Cover design by Russell Richardson

Set in size of 12pt and Minion Pro by Manila Typesetting Company, Makati, Philippines

Printed in the USA

10 9 8 7 6 5 4 3 2 1

Contents

Preface xv

Acknowledgments xvii

1 Introduction to Bio-Inspired Devices: Design, Manufacture, and Application 1

1.1 Understanding Bio-Inspiration 2
1.2 Historical Context and Evolution of Bio-Inspired Design 5
1.3 Biological Structures and Mechanisms as Models 7
1.4 Principles of Bio-Inspired Design 9
1.5 Bio-Inspired Materials and Fabrication Techniques 13
1.6 Bio-Inspired Devices in Engineering Systems 16
1.7 Bio-Inspired Devices in Medicine and Healthcare 20
1.8 Bio-Inspired Devices in Environmental and Agricultural Applications 23
1.9 Bio-Inspired Robotics and Autonomous Systems 25
1.10 Conclusion, Future Trends, and Challenges 27
References 27

2 Bio-Inspired Manufacturing and Fabrication Processes 31

2.1 Introduction 32
2.2 Biological Principles Used in Bio-Inspired Manufacturing 36
2.3 Bio-Inspired Fabrication Strategies and Technologies 38
2.3.1 Additive Manufacturing 38
2.3.2 Functionally Graded Materials Manufacturing 40
2.3.3 Lithography as a Manufacturing Process 41
2.3.4 Self-Assembly and Directed Self-Assembly 42
2.3.5 Bioprinting 44
2.3.6 Microfluidics as a Fabrication of Bio-Inspired Systems 45
2.3.7 Bio Mineralization 47
2.3.8 Origami-Inspired Design 48

2.3.9 Bioelectronic Interfaces 49
2.4 Materials and Material Systems in Bio-Inspired Manufacturing 50
2.5 Conclusion 52
References 53

3 Bio-Inspired Computation: Leveraging Ant Behavior for Smart Algorithm Design 57
3.1 Introduction 58
3.2 Algorithms 59
3.2.1 Path Planning Algorithms 59
3.2.2 Ant System 60
3.2.3 MAX–MIN Ant System 61
3.2.4 AddACO 62
3.3 Swarm Intelligence Algorithms 63
3.3.1 Aphid-Ant Mutualism 63
3.3.2 Decentralized Control Algorithm for a Collection of Robots 65
3.4 Algorithm for Collective Sorting Robots 67
3.4.1 Simulation Environment and Algorithm 68
3.4.2 Clustering and Sorting Process 68
3.5 Search Algorithms 69
3.5.1 Ant Walk Algorithm 69
3.5.2 Hidden Markov Model 70
3.6 Link Prediction Algorithm 72
3.6.1 Quantum-Based Optimization Algorithm 72
3.7 Application of Ant-Inspired Algorithms 73
3.7.1 Internet of Drones 73
3.7.1.1 Problem Definition 74
3.7.1.2 Key Aspects of the Algorithm 74
3.8 Dynamic Task Allocation Problem – Machine Scheduling 76
3.8.1 Problem Definition 76
3.8.2 Key Aspects of the Algorithm 77
3.9 Additive Manufacturing Tool Path Optimization Using Ant Colony Optimization 79
3.9.1 Methodology 80
3.10 Conclusions 80
3.11 Future Research Opportunities 81
References 82

4 Bio-Inspired Fabrication for Sustainable Water Scarcity Solutions: Learning from Nature 85
4.1 Introduction 86
4.2 Inspiration from Beetles 86
4.3 Water Harvesting Process 88
4.4 Fabrication of Beetle-Inspired Surfaces 88
4.4.1 Overview of Fabrication Techniques 88
4.5 Materials for Water Harvesting 92
4.6 Challenges 92
4.7 Mechanism of Water Harvesting 93
4.7.1 Interaction Between Hydrophilic-Hydrophobic Surfaces 93
4.7.2 Wetting Models 94
4.7.2.1 Wenzel Model 94
4.7.2.2 Cassie–Baxter Model 95
4.7.2.3 Dynamic Wetting 95
4.8 Comparison with Other Natural Systems 95
4.9 Mathematical Modeling and Analysis 96
4.9.1 Formation of Droplet and Related Surface Energy 96
4.9.2 Dynamics of Growth of Drops and Collection 97
4.9.3 Contact Angle and Surface Energy 98
4.9.4 Performance Optimization Models 99
4.9.5 Scalability Models 100
4.10 Applications Based on Water-Harvesting Mechanisms of the Beetle 101
4.11 Conclusion 103
4.12 Future Research Challenges and Opportunities 103
4.12.1 Adaptive Surface Technologies 104
4.12.2 Advancements in Manufacturing 104
References 105

5 Biomimetic Bone Implant Design: Materials, Manufacturing, and Modeling 109
5.1 Introduction 110
5.2 Biomimetic Bone Implants 110
5.2.1 Materials for Biomimetic Bone Implant 111
5.2.1.1 Classical Materials and Their Deficiencies 111
5.2.1.2 New Developments and Hybrid Materials 111
5.3 Fabrication Methods 111
5.3.1 Additive Manufacturing 112
5.3.2 Other Techniques 112

5.4 Finite Element Modeling and Stress Analysis 112
5.4.1 Finite Element Modeling in Implant Design 112
5.4.2 Practical Applications and Benefits 112
5.5 Materials Design for Bone Implants 113
5.5.1 Criteria for Material Selection 113
5.5.1.1 Mechanical Strength and Load-Bearing Capacity 113
5.5.1.2 Biocompatibility and Osteointegration 114
5.5.1.3 Bioactivity and Osteoinductivity 114
5.5.2 Degradability and Controlled Resorption 116
5.6 Materials Combinations for Implant 116
5.6.1 Titanium and Titanium Alloys 117
5.6.2 Hydroxyapatite and Bioglass 118
5.6.3 Polymers (PCL and PLA) 118
5.6.4 Composite Materials 118
5.6.5 Smart Materials and Functionalized Scaffolds 119
5.7 Fabrication Techniques 121
5.7.1 Additive Manufacturing 121
5.7.1.1 Key Attributes of Additive Manufacturing 121
5.7.1.2 Applications 122
5.7.1.3 Limitations 122
5.7.2 Traditional Techniques 122
5.8 Stress Analysis and Finite Element Modeling Simulation 123
5.8.1 Stress Analysis Role in Implants 123
5.9 Mathematical Modeling and Analysis 124
5.9.1 Mechanical Behavior and Stress Analysis 124
5.9.1.1 Fundamental Stress-Strain Relationships 124
5.9.1.2 Bending Stress in Cylindrical Implants 125
5.9.1.3 Stress Distribution in Porous Scaffolds 125
5.9.1.4 Yield Strength in Implants 126
5.9.2 Thermal and Fabrication Modeling 126
5.9.2.1 Heat Conduction in Scaffold Materials 127
5.9.2.2 Thermal Stress in Implants 127
5.9.3 Biological and Remodeling Models 128
5.9.3.1 Bone Adaptation Under Stress 128
5.9.3.2 Cellular Growth in Scaffolds 128
5.10 Porosity Design and Optimization 129
5.10.1 Level-Set Equations for Porosity Control 129

5.11 Conclusion 130
5.12 Scope for Future Work 130
References 131

6 Gecko-Inspired Adhesives: A Multiscale Approach to Biomimetic Adhesion and Applications 135
6.1 Introduction 136
6.2 Historical Background 136
6.3 Scientific Relevance of Gecko Adhesion 137
6.4 Significance of Biomimetic Studies 137
6.5 Objective of the Review 138
6.6 Structure-Function Relationship 138
6.7 Mechanisms Underpinning Gecko Adhesion 139
6.8 Fabrication Techniques for Gecko-Inspired Adhesives 140
6.9 Potential Applications 140
6.10 Biology of Gecko Feet 141
6.10.1 Structure of Gecko Feet 142
6.10.2 Role of Setae and Spatulae 142
6.10.3 The Hierarchical Structure: Setae and Spatulae 142
6.10.4 Material and Mechanical Properties of Setae 143
6.10.5 Adaptations for Environmental Variability 143
6.10.6 Role of Claws in Complementing Adhesion 143
6.10.7 Evolutionary Perspective on Adhesive Morphology 144
6.11 Mechanism of Adhesion 144
6.11.1 Van der Waals Forces 144
6.11.2 Hierarchical Structure of Gecko Feet 144
6.11.3 Contact Optimization 145
6.11.4 Detachment Mechanism 145
6.12 Contributions of Friction to Adhesive Performance 145
6.12.1 Impact of Surface Energy and Wettability 146
6.12.2 Geometric Contributions to Adhesion 146
6.12.3 Interplay of Forces and Structural Features 146
6.12.4 Influence of Environmental Factors 147
6.12.5 Technological Implications and Future Directions 147
6.12.6 Directional and Reversible Adhesion 148
6.12.7 Anisotropy in Adhesive Forces 148
6.13 Mechanisms Underpinning Reversibility 148
6.13.1 Functional Role of Load and Shear Forces 149
6.13.2 The Role of Structural Hierarchy in Directionality 149

6.13.3 Environmental Influence on Directional and Reversible Adhesion 150
6.13.4 Evolutionary Adaptations for Directionality and Reversibility 150
6.13.5 Applications of Directional and Reversible Adhesion in Biomimetics 151
6.13.6 Integration of Mechanisms in Locomotion and Adaptation 151
6.14 Materials and Design Principles in Gecko-Inspired Adhesives 151
6.14.1 Material Selection 152
6.14.2 Structural Design Principles 152
6.14.3 Key Design Properties 153
6.15 Production Methods 153
6.16 Applications of Gecko-Inspired Adhesives 153
6.16.1 Soft Robotic Gripper with Gecko-Inspired Adhesive 153
6.16.2 Robotics and Automation 154
6.16.3 Applications in Robotics and Space Exploration 154
6.16.4 Advancements in Medical Adhesives 155
6.17 Future Directions in Gecko Adhesion Research and Applications 155
6.17.1 Expansion of Knowledge of Adhesion Processes at Molecular Dimensions 156
6.17.2 Investigating Novel Features Derived through Evolution in Different Organisms 156
6.18 Conclusion 157
References 158

7 Manufacturing Shark Skin-Inspired Swimwear with Enhanced Fluid Dynamics and Performance 163
7.1 Introduction 163
7.2 Morphological, Functional, and Hydrodynamic Properties of Shark Skin 166
7.3 Fabrication, Applications, and Hydrodynamic Properties of Biomimetic Shark Skin 169
7.4 Manufacturing Techniques of Shark Skin-Patterned Fabrics 173
7.4.1 Digital Modeling and Computer-Aided Design 173
7.4.2 Use of Additive Manufacturing for the Fabrication of Skin 175
7.4.3 Nano- and Microstructuring 177

7.4.4 Integration with Flexible Substrates 177
7.4.5 Coating Techniques 178
7.5 Sensors and Smart Sensors Used in Sharkskin Fabrics 179
7.6 Actuators Used in Sharkskin Fabrics 180
7.7 Conclusions 181
7.8 Future Research Opportunities 182
References 183

8 Spider Silk-Inspired Ropes and Fibers: From Nature's Design to Artificial Production 185
8.1 Introduction 185
8.2 Primary and Secondary Structures of Spider Silk 187
8.3 Orb-Weaver Spider and Dragline Silk 188
8.4 Major Ampullate Silk 189
8.5 Nanofibrillar Structure 191
8.6 Skin Core Structure of Dragline Silk 192
8.7 Factors Affecting the Mechanical Properties of Spider Silk 192
8.7.1 Reeling Speed 192
8.7.2 Heat and pH 193
8.7.3 Shear Force During Spinning 193
8.7.4 Some Environmental Factors 193
8.7.4.1 Wind 193
8.7.4.2 Temperature 194
8.7.4.3 Humidity 194
8.7.4.4 Exposure to UV Radiation 194
8.8 Raw Materials 195
8.8.1 Recombinant Spidroins 195
8.8.2 Non-Spidroin Proteins and Polypeptides 196
8.8.3 Polymer Materials 197
8.9 Methods of Fiber Production 197
8.9.1 Dry Spinning 197
8.9.2 Wet Spinning 198
8.9.3 Electrospinning 199
8.10 Microfluidic Spinning 200
8.10.1 Novel Spinning Methods 200
8.11 Physical Properties of Spider Silk and Inspired Artificial Fibers 200
8.11.1 Water Response 201
8.11.2 Light Transmission 201

8.11.3 Torsional Energy Dissipation 201
8.11.4 Thermal Conductance 202
8.12 Applications of Spider Silk-Inspired Fibers 202
8.12.1 Construction and Civil Engineering 202
8.12.2 Electronics and Energy Transmission 203
8.12.3 Robotics 203
8.12.4 Aerospace 203
8.12.5 Agriculture 204
8.12.6 Other Applications 204
8.13 Conclusions and Future Scope 205
References 206

9 Biomimetic Design and Fabrication of Hummingbird-Like Micro Unmanned Aerial Vehicles 209
9.1 Introduction 210
9.2 Importance of Hummingbird-Inspired Drones 211
9.3 Relevance of Hummingbird-Inspired Drones 212
9.4 Evolution of Bio-Inspired Drones 213
9.5 Hummingbird Flight Mechanics 214
9.5.1 Calibration of Sensors and Error Correction 215
9.5.2 Flight Time and Path Planning 215
9.5.3 Localization and Tracking 216
9.6 Characteristics of Hummingbirds 216
9.7 How Do Hovering Hummingbirds Work 217
9.8 Materials and Fabrication Techniques 218
9.9 Application and Potential Uses 220
9.9.1 Surveillance and Reconnaissance 220
9.9.2 Medical and Resource Operations 220
9.9.3 Environmental and Ethical Considerations 221
9.9.4 Unmanned Aerial Vehicles for Air Monitoring 221
9.9.5 Unmanned Aerial Vehicles for Land Monitoring 222
9.9.6 Unmanned Aerial Vehicles for Marine and Water Monitoring 222
9.10 Regulations and Policy for Operating Unmanned Aerial Vehicles 223
9.11 Limitations and Challenges 223
9.12 Current Challenges and Solutions 224
9.12.1 Manufacturing Flapping Wing Mechanism 224
9.12.2 Grasping During Flight 224
9.12.3 Damage Mitigation 227

9.12.4 Collision Avoidance 227
9.12.5 Flight Efficiency 228
9.13 Conclusions 232
References 233

10 Biomimicry in Automotive Design: Nature-Inspired Innovations for Efficiency, Performance, and Sustainability 239
10.1 Introduction 240
10.2 Classification of Bio-Inspired Designs in Automobiles 241
10.2.1 Aerodynamic Drag Reduction Applications 241
10.2.1.1 Biomimicry and Aerodynamics 243
10.2.1.2 Streamlined Shapes 244
10.2.1.3 Surface Textures 245
10.2.1.4 Adaptive Aerodynamics 245
10.2.2 Structural Applications 248
10.2.3 Wheel Assembly Applications 255
10.2.4 Powertrain Applications 258
10.2.5 Applications in Digital Manufacturing 259
10.3 Challenges in Adopting Bio-Inspired Designs for Automobiles 261
10.3.1 Complexity of Nature's Designs 261
10.3.2 Material Limitations 262
10.3.3 Technological and Engineering Constraints 262
10.3.4 Manufacturing and Production 262
10.3.5 Cost of Research and Development 263
10.3.6 Sustainability Concerns 263
10.3.7 Regulatory and Safety Compliance 264
10.4 Conclusions 265
10.5 Future Research Opportunities 266
Bibliography 267

Index 271

Preface

Based on the macro-to-nanofabrication of bioengineered materials, spectroscopic characterization, environmentally friendly production pathways, and sustainable industrial-scale applications, this volume examines recent breakthroughs in innovative research across inter- and cross-disciplinary sciences. It comprises ten chapters covering a broad range of topics, including bio-inspired devices, biomimetic bone implant design, gecko-inspired adhesives, micro–unmanned aerial vehicles, and biomimicry in automotive engineering.

Over the past few decades, significant attention has been devoted to the self-assembly of structural proteins, which form intricate hierarchical materials combining exceptional mechanical, chemical, and transport properties. This highly regulated process spans both nano- and macroscales, making such materials especially attractive for the development of integrated, adaptive systems in bio-inspired manufacturing. Parallel advances in materials development and their application in bio-inspired devices have further accelerated progress in this field.

This volume is organized into ten chapters.

Chapter 1 introduces bio-design, tracing its historical context while outlining unique biological structures, mechanisms, materials, fundamental principles, fabrication techniques, and applications that have enhanced human life. Chapter 2 focuses on material considerations and advanced fabrication technologies—including additive manufacturing, soft lithography, photolithography, bioprinting, electrospinning, laser-based processes, and protein-based techniques—offering detailed comparisons in terms of efficiency, sustainability, adaptability, energy use, and customization. The integration of biological principles with diverse engineering disciplines is shown to enable innovative technologies, products, devices, and systems for future applications.

Chapter 3 explores algorithms inspired by ant behavior, emphasizing ant colony optimization and broader swarm intelligence approaches that abstract essential behavioral characteristics rather than replicating biological systems exactly. Chapter 4 critically examines bio-inspired technologies for biosensing and fog harvesting, highlighting sustainable and environmentally friendly solutions to global water scarcity. Chapter 5 integrates materials science, advanced manufacturing, and computational modeling to provide an overview of current and emerging biomimetic bone implants, underscoring the need to balance biological integration with mechanical performance in regenerative medicine.

Chapter 6 reviews bio-inspired adhesive technologies and their applications in robotics, medical devices, and wearable systems, including gecko-inspired dry adhesives that enable climbing robots, enhanced soft robotic grippers, and improved medical adhesives for wound closure and drug delivery. Evolutionary and ecological perspectives on gecko adhesion are also discussed. Chapter 7 examines sharkskin-inspired materials, focusing on denticle mechanics and their role in enhancing aerodynamic performance, along with future research directions such as 3D-printed sharkskin-based swimwear.

Chapter 8 discusses the properties of spider silk and its potential applications, particularly as a high-toughness material capable of resisting fracture. Chapter 9 presents an overview of the design and manufacturing principles of nature-inspired miniature UAVs, demonstrating how biological forms and flight mechanisms inform advanced engineering solutions. Chapter 10 highlights the growing role of biomimicry in automotive design, showcasing innovations inspired by animals and plants that improve aerodynamics, reduce emissions, and enhance fuel efficiency.

Taken together, the chapters demonstrate that bio-inspired manufacturing systems enabled by advanced fabrication methods are increasingly moving toward scalable implementation. As biomaterials continue to evolve and become more cost-effective, they hold the potential to replace environmentally harmful fossil-based materials and significantly advance biomedical and industrial performance. This volume synthesizes recent developments that point toward future research directions in customized bio-inspired devices, contributing to the broader goal of affordable, efficient, and sustainable production across academia, research, and industry.

The Editors
January 2026

Acknowledgements

We extend our sincere gratitude to everyone who helped to create this book. Its realization and completion would not have been possible without the guidance, support, and encouragement of many individuals and organizations.

We would like to extend special thanks to our family and friends for their patience, understanding, and unwavering support throughout this demanding journey. Their encouragement helped us navigate both the challenges and rewards of the research process.

1

Introduction to Bio-Inspired Devices: Design, Manufacture, and Application

Abstract

Bio-inspired design and manufacturing is a multidisciplinary domain that requires biologists, engineers, doctors, chemists, environmentalists, artists, and many more. Nature provides everything essential for humans to exist on this planet. The materials, processes, structures, mechanisms, and behaviors of living beings provide great insights for researchers to mimic nature and produce devices, products, and machinery for the betterment of mankind. Most of the available biodesigns in nature are eco-friendly, sustainable, cost-effective, and use the principle of minimum parts and minimum assembly. Numerous examples in nature can be mimicked. Micro and nano structures on beetles, used for water harvesting in the desert, spider's vibration ability, nacre mortar and brick structure, shark skin superstructure, aerodynamics of birds, flexibility of the trunk of an elephant, broad eyes of a frog, and the flexibility of an octopus are just a few examples. There are many optimization principles hidden in the lives of ants, bees, termites, etc. Researchers can realize bio-inspired design into a useful product that has improved many phases of our lives. Advanced technologies based on laser processing, additive manufacturing, and electrospinning are available to fabricate bio-inspired products. This chapter introduces biodesign, its historical context, unique structures and mechanisms, materials, basic principles, and various fabrication techniques to prove that bio-inspired products also improve many applications and human life.

***Keywords*:** Bio-inspired design, sustainability, efficiency, smart materials, interdisciplinary approach

Jagadeesha T. and Sandip Kunar. Bio-Inspired Manufacturing Systems, (1–30) © 2026 Scrivener Publishing LLC

1.1 Understanding Bio-Inspiration

Bio-inspiration means observing nature and biological systems around us and studying the complex behavior of those systems and mimicking or duplicating those processes or materials, designs, or manufacturing in engineering to produce products that are eco-friendly, productive, cost-effective, sustainable, and environmentally friendly. A deeper understanding is essential to imitate nature, functions, and structure to arrive at cost-effective design solutions. Fundamental physics of nature must be understood first to apply those natural rules and laws in producing a product of useful nature for human beings [1]. In this creation, millions of biological organisms live in varied atmospheres ranging from water to air, from harsh to very harsh atmospheres, from high temperatures to the lowest temperatures, and so on. Nature not only gives great ideas but also provides a fundamental framework in the design of automobiles, aircraft, robotics, manufacturing, processes, medicine, and so on. A major advantage of a bio-inspired system is efficiency. Nature always uses principles of minimum energy and minimum waste. Their efficiency is due to their ability to reuse. A simple example of aircraft design is inspired by a bird, which has two wings and very complex maneuverability to adjust the flow so that drag is minimized, landing is smooth, and take off is without any extra effort. Swinging is done to maintain a variable temperature [2]. A bird is a perfect example of perfect maneuverability, enhanced resilience, and very efficient systems. Nature works both at the microscopic level and the macroscopic level, and humans have an opportunity to learn from both classes of creation. Nature built at the microscopic level is highly complex, self-healing, and adaptive in nature. Studies of many microscopic organisms have led to several innovations like smart materials (magnetorheological fluid, electrorheological fluids, shape memory alloys, meta materials, self-healing materials, self-aligning materials). Complex structures of lotus leaf, which are responsible for self-cleaning nanoscale hairy structures with gradients on the geckoes, are just a few examples.

Many species on Earth have changed due to evolutionary processes, and some have totally vanished from the Earth due to calamities and natural phenomena, some of which are unknown. Nature has solved many complex, optimized, and time-tested problems. This evolution of nature, animals, plants, insects, micro- and macro-organisms provides wide ideas for engineering analysis and the construction of devices and machinery. It provides alternate designs, which are evolutionary. Although nature appears simple, in reality, it is a complex interdisciplinary hub of biology, chemistry, physics, engineering, and modeling. Physicists and chemists study nature and capture the essence of biological creation, and they provide the fundamental laws and science behind the behavior of animals and plants [3]. An engineer translates science into technology for better human comfort. Computational people use algorithms and optimize the technology or processes. Even the optimization algorithms are also nature-inspired, like ants, swarms, and bees. There are many branches of engineering, which are bio-inspired, like bio robotics, bio mechatronics, biometrics, and bio fabrication, to name a few. Sustainability is the very design of bio-inspired design. Most designers who design conventional manufacturing or processes have worked on performance and savings. But nature always devised environmentally friendly systems. Therefore, it is essential to design and manufacture a bio-inspired system that works hand in hand with nature. Designers should use nontoxic, biodegradable, renewable, reusable, and recyclable materials, which are not only sustainable but also give peace to human beings.

Designers often encounter constraints and material selection; manufacturing processes do have constraints. Most man-made structures are rigid, but nature has produced complex, compliant, and flexible structures. For example, an octopus arm is highly flexible; elephant trunks are flexible structures; the fins of sealions; fish show high flexibility and compliance. The flexibility of an octopus is mimicked in the soft grippers of robotics systems [4]. Nature's complexity can only be understood when internal complexity is given due attention rather than only the surface phenomenon. Today,

Table 1.1 Different aspects of bio-inspired manufacturing systems.

Sl. no.	Aspects/terminology	Explanation
1	Definition	Studying with deep insight with curiosity, wonder, and mimicking nature's materials, structures, processes, or designs to develop eco-friendly, cost-effective, and sustainable solutions.
2	Importance	Deep understanding of physics, chemistry, and the structure existing in nature for replication for betterment.
3	Environment Adaptability	Water and air can exist in different forms and under different conditions and places, and it is an inspiration for innumerable applications in automotive, medicine, aerospace, etc.
4	Efficiency Principle	Nature always uses the principle of minimum energy and zero waste.
5	Microscopic Inspiration	Many structures are possible in organisms and provide inspiration.
6	Natural Structures	Lotus leaf (self-cleaning), Gecko foot (nanoscale hairs).
7	Evolutionary Learning	Ants, bees, and termites provide great lessons for evolutionary learning.
8	Interdisciplinary Nature	Bio-inspiration combines all biology, chemistry, physics, engineering, modeling, and computing fields. It is a multidisciplinary domain.
9	Branches of Bioengineering	Bio-robotics, bio-mechatronics, biometrics, bio-fabrication, etc.
10	Sustainability Focus	Biodegradability, recyclability, reusability, eco-conscious materials, and processes are the strengths of bio-inspired designs.
11	Flexibility & Compliance	Nature provides flexible and compliant structures, such as elephant trunks, octopus arms, and fish fins.

(*Continued*)

Table 1.1 Different aspects of bio-inspired manufacturing systems. (*Continued*)

Sl. no.	Aspects/terminology	Explanation
12	Technology Aids	Additive manufacturing, electrospinning, and many more are available for the fabrication of bio-inspired products.
13	Global Challenges	Water scarcity, energy conservation, and environmental sustainability can be solved using bio-inspired designs.

many structures can be mimicked using a 3D scanner, 3D printing, MR imaging, computer-aided design packages, analysis packages, simulation packages, and so on. Many creative designs are possible with bio-mimicry design by observing nature's solution to problems, be it a water harvesting by a beetle, or a complex web by a spider. To develop efficient, cost-effective, sustainable, environmentally friendly devices, processes, and systems, one needs to have a good understanding of bio-inspiration. Many global issues, like water harvesting, climate change, and energy conservation, can be solved by studying nature and getting inspiration to design devices for better human comforts. General terminologies used in bio-inspired design, manufacture, and material are summarized in Table 1.1.

1.2 Historical Context and Evolution of Bio-Inspired Design

Civilization has evolved by observing and mimicking nature. Human beings observed the killing of one animal by another and devised sharp-edged tools, which mimic the nails of tigers or lions. When raining, animals take shelter in caves. Through mere observation, humans built primitive houses. Humans observed how birds build their nests and mimicked them. Aircraft design was inspired by birds. Ship design was inspired by fish anatomy and behavior. Submarines were developed by observing the tortoise, which can survive both on land and in water. Initially, human beings copied nature for survival and basic needs. Later, humans mimicked nature for more comfort, like air conditioning, automobiles, etc. Earlier humans solved many problems by simple observation and deep understanding of nature. There were no complex computers available yet; remarkable progress was made by early civilizations through intuitive awareness and consciousness of their surroundings. Da Vinci studied the nature of birds and bats in detail with such precision and in-depth observation; he could create flying machines. He could come up with a precise aerodynamics theory by observing the fluid flow over birds' wings and how feathers offer a complex and compliant mechanism to adjust lift and drag to give a perfect aerodynamic pressure and velocity [5]. Although Da Vinci did it artfully, it later led to functional design.

The Industrial Revolution and war both brought out many advanced technologies to mankind, both destructive and constructive. Industrialists, bureaucrats, and technocrats used the concept of the reductionist principle to produce components in large quantities through the principle of standardization. Flexible, adaptable designs were developed during the Industrial Revolution with an emphasis on efficient use of resources. One of the hallmarks was the development of biomimetics. In 1950, Otto Schmitt developed an electronics circuit using the bio mimics of the squid's nervous system. Jack Steele coined the new word called bionics, which integrated both biological functions and engineering functions. The invention of electron microscopy opened the door for a large number of inventions because many biological details, structures from the nano level to the micro level, became accessible to mankind. It became easy for scientists and engineers to make a correct assessment about the biological features, mechanisms, and behavior based on what they saw under the microscope [6]. Micro and nanotechnologies were developed during these periods. Velcro was developed by engineers with inspiration from the burdock plant, and many practical inventions were realised by observation of nature and putting it into engineering the components. During the mid-20th century, architects designed buildings that could regulate the temperature passively by mimicking termite mounds. Several coating materials have been developed based on the nanostructure of lotus leaves, which have a self-cleaning ability. Now in the 21st century, scientists and engineers can model, simulate, and manufacture complex components and devices to make soft robotics, bio-mimicked electronics, etc.

1.3 Biological Structures and Mechanisms as Models

Structures and mechanisms are the building blocks of any product or device. Structure is needed to support the system, and a mechanism is needed to perform a specific task. Nature has innumerable structures. They range from organelles to animal appendages and plants. Observation, understanding of these structures can give a product of innovative design and sustainable manufacturing. For example, limbs of quadrupeds, birds' wings, fins of sharks, and suction cups of cephalopods inspired many innovative devices both in the engineering and medical fields. Biological structures offer flexibility, compliance, and stiffness. Many grades' structures can be created by mimicking the microscopic arrangement of structures in animals and plant life [7]. Scientists have developed a micro air level by observing drag, lift, and fluid flow across the structures of dragonflies. Micro air level works in harsh environments, offering stable maneuverability and other improved aerodynamic properties, such as drag and lift, even in constrained spaces. Interesting bio-inspiration lies in the multi-scale structure in nature. There are many organisms or parts of organisms where one can observe and implement multiple structures in their design to improve overall efficiency. Bone is a multi-scale structured part in both animals and humans. It is a composite material made of collagen and hydroxyapatite. The scale ranges from micro to nano scale. These graded structures provide good toughness, lightness, and strength to bone. Many multiscale structures are produced by engineers by mimicking the bone structure. Another example is a multi-structure observed in nacre. It is a biopolymer structure with a brick-and-mortar structure [8]. This structure is fracture-resistant and can dissipate energy easily, thereby maintaining the temperature. Many body armor-resistant materials are built by mimicking the nacre structure. Thus, nature provides the basic building blocks for structures.

Biological surfaces and curvatures, micro, nano, and gradient surfaces provide significant input to designers in terms of functions. Laser texturing, electrospinning methods are developed by mimicking animal skin, such as shark, fish, and surface properties like wet angles, adhesion, and friction behavior. Wear resistance can be mimicked to get super hydrophobic or hydrophilic surfaces [9]. Van der Walls' forces for adhesion, gecko feet, provide greater insight into the production of wet and dry adhesives without mechanical fastening like bolts, nuts, rivets, and welding. Many manufacturing processes can be either eliminated or minimized, or modified using many of the bio-inspired observations and implementations, both at the microscopic and macroscopic levels. Many advanced mechanisms can be found in animals, insects, and plants. Self-healing mechanisms, camouflaging, and advanced sensor mechanisms are just a few examples to illustrate responsive and highly intelligent, and evolved systems exist in nature, which can be mimicked not only to produce novel materials like responsive textiles but also new mechanisms. Nature uses the principle of minimum parts, minimum energy, highly efficient systems, resilient and adaptable mechanisms [10]. Many advanced branches of engineering, like transportation, robotics, automation, data processing, and control engineering, use bio-inspired mechanisms to optimize the system's performance and to improve productivity. Many optimum algorithms are generated by observing ants, swarms, and the flocking dynamic behavior of birds and bees. Excellent storage and retrieval systems. Data processing and logistics can be improved by studying ants and bees.

Today, multidisciplinary knowledge and applications have yielded many products useful to mankind. Biology, medicine, and engineering are needed to design and manufacture advanced medical devices, such as MR scans and many varieties of bio-hybrid devices or systems. Biologists are needed to understand genetic organization, protein formation, and metabolism. Doctors need to give an expert data analysis of protein folding, microorganism behavior, healing mechanisms, and blood behavior, etc. Engineers are needed to convert science into technology, processes, devices, machines, robots, etc., for the betterment of mankind [11]. All types of engineers, like mechanical, electrical, electronics, communication, information technology, artificial intelligence, and machine intelligence, are needed to make an economic and useful product. In the past centuries, there have been groundbreaking innovations in all fields of engineering, medicine, and biology, ranging from architecture, nuclear, aerospace, medicine, consumer products, and many more.

1.4 Principles of Bio-Inspired Design

Natural principles hidden in nature, for example, gravity, provide a profound guideline to biologists, doctors, engineers, and many more to convert the natural phenomenon into innovations, products, processes, and machinery. Nature adapts, evolves, and grows in a very systematic, simple, and profound way. Not just copying, imitation helps; rather, deep and profound thinking is needed to understand the bio-inspired processes or devices [12]. As discussed earlier, a multidisciplinary design approach is essential to learn and apply the essential principles from nature. Nature is good at combining one or more things. For example, a form and function combination. Structures are built to do certain advanced functions. Nature neither wastes material nor energy. A close observation of nature reveals that curves, structures, specific functions, and colors are built into the system, either to optimize performance or to give maximum efficiency. The camouflage of animals is exceptional and is used in military applications to blend in with the surroundings so that the enemy will not spot them. The mechanism of change of body color to blend in with nature is quite complex, but it can be easily incorporated by the military by proper choice of color, even more challenging if response textiles are used by military personnel [13]. In nature, there are complex mechanisms that are easy to implement, and simple phenomena, like gravity, that are difficult to implement in a laboratory. Another mechanism is the shape of a fish or bird; their structures are superior from an aerodynamic and hydrodynamic point of view and are used in vehicles, submarines, and sports cars.

Many animals have parts of the body that can do multiple functions. For example, human skin has multiple functions to play. It acts as a protective layer; it is a touch sensor and actuator (hand or leg), and regulates the body temperature. Its elastic, plastic, fatigue, and impact behavior offer several multifunctional devices. Space engineers have successfully designed multifunctional materials that are self-healing, can carry loads, have sensing abilities, and actuator abilities [14]. There are many classes of Piezoelectric, thermoelectric, and metamaterials developed for medical applications. When functions are combined, the overall cost can be reduced, and the lead time can be reduced. Strength, flexibility, stiffness, and weight reduction can be improved by using and developing a hierarchical structure. Engineers have explored layered structures, graded structures in terms of micro to nanoscale or vice versa, and gradient structures in terms of materials by mimicking bone structure, wood, and silk. By changing microscopic properties, it is shown that macroscopic properties can be changed, and by changing macroscopic properties, one can change the microscopic properties. Heat treatment, alloying, and additive manufacturing for complex shapes and sizes are good examples.

Flexibility, responsiveness, and adaptability are key to animals' and plants' survival. Solar reflectors are based on the simple concept of plants' leaves, which can orient for optimum sunlight for photosynthesis. There are many animals, insects, and plants in nature, which can change to input stimuli such as light, temperature, pressure, touch, or chemical signals. Some animals use chemicals for their survival and prey, such as snakes and kamado dragons. Smart response textiles, soft robots, and many medical habitation devices are possible by using stimulus-response materials. Adaptive building facades, which are bioinspired designs, can optimize performance in real-time dynamically. Many supply chain managers and data analysts use today's concept of decentralized control, inspired by ants and birds. These work with local instructions without central commands to locally maximize to achieve global outcomes, which could be either maximization or minimization. Robots, dependable, reliable, scalable, and sustainable algorithms are computerized to save cost, effort, time, and traffic. This will ultimately bring more customer satisfaction and correct delivery, and thus more profit. Amazon online shopping is the best example of decentralized supply chain management systems, like an ant colony.

Any successful design, product, device, or system must be energy efficient. Energy-efficient system uses minimum energy input and waste, or loss is minimal, so output is either optimized or maximum. Most biological systems work with passive mechanisms with optimum topology, geometry, and surface energy, and create low-energy-consuming systems. Many of these create optimized wind, heat, and light, which inspired the manufacture of turbines, submarines, and other underwater equipment. Any technology, either bio-inspired or man-made, must be sustainable. Biocycle nature creates usually closed-loop systems. Many of the byproducts of plants are biodegradable and have multiple functions, even after the crops, like the banana tree. So, most plants' lives are self-sustainable; they help nature complete a closed loop by becoming a source for another organism, plant, or animal. Biological cycle, biogeochemical cycles, fungal cycles, and algae are a few examples of closed life cycles [15]. They are eco-friendly, suitable, and regenerative technologies with minimal waste.

Any design or system must be robust, adaptable, and scalable to function well under various conditions, either process or machine-dependent. Many of the systems are now modular systems in DNC and CNC machines. These are inspired by a spider's web. Spiders can still work even if part of the web is destroyed and replace the spoiled or destroyed one in a modular way. The same concept is applied in computer numerical control machines, where control systems are modular; even if one module fails, it can still work, of course, with less efficiency. When the module is replaced, it again picks up the speed and works with robustness and good scalability. Built-in software is also adoptable, and the structure is also resilient against fatigue and impact loads, which arise from severe cutting conditions. Fail-safe and safe fail are two important mechanisms that are built into nature. Any devices, processes, or mechanisms that fail should fail safely without destroying others or human beings. And, if the device or mechanism fails, it should have interlocks so that it does not cause harm to others. These are achieved using advanced sensor technology, monitoring systems, AI tools, and built-in backup systems by building redundancy and fault-tolerant systems or damage-tolerant systems. Redundancy is very important in nuclear power plants from a safety point of view. Multiple redundancy is used when catastrophic failure directly affects human life or safety.

Biological design is quite often undergoing many iterations to adopt the best possible design. Birds build a nest with a two-door mechanism, one that is empty where predators cannot find prey, and another secret door through which only the bird can go and lay eggs. Birds evolved their nest's structure based on other conditions, and this type of evolutionary optimization is the underlying principle for their survival. Many non-intuitive designs can be generated by using AI techniques today [16]. Many biological systems consider their operating environment as a complement and not a competitor. contextuality and harmony are natural to many plants, animals, and organisms. Many organisms depend upon other habitats and adapt to new environments through migration for better weather conditions and survival. Termites, for example, build tall and very complex mounds that can maintain local temperature and humidity. The termites build a central chimney and side walls for excellent ventilation and use a stack so that hot air goes out and cold air comes in due to heat transfer through convection. Such a complex tunnelling has inspired many architects to adopt the principle of contextuality to build a structure that can adjust to local climates and ventilation systems, which are stack structures and self-maintainable and sustainable through unconventional energy like wind, solar, and water. The holistic approach nature follows is worth imitating and adopting.

All these principles described above are not applied individually but rather holistically in nature. That is the reason why nature can produce highly adaptable, efficient, sustainable designs. Many materials, including stimulus-response material, are designed for the soft hand of a robot system by mimicking the octopus, which is multi-functional, adaptable, flexible, energy-efficient, and robust. Nature provides an abstract and subtle model, which can be translated into a functional model within the boundaries of engineering, medicine, biology, and other fields.

1.5 Bio-Inspired Materials and Fabrication Techniques

The selection of material and fabrication techniques to realize the design is as important as the design process itself. Be it an engineering product, consumer product, medical device, military equipment, or machinery, materials and fabrication are extremely important to give cost benefits and efficiency at the device level, as well as the system level. Nature has developed high-performance materials such as composites and provides a very sustainable and efficient manufacturing such as leaves of coconuts. Nature always offers inspiration to develop novel material and manufacturing processes to suit both cost and performance in terms of function and delivery. There are many biological materials, biopolymers such as nacre, spider silk, cellulose from plants, bones, and leaves that give good, combined properties such as toughness, light weight, stiffness, and resilience. Chemical composition, combined with a graded structure and material, gives such an exceptional property. Structures range from macro to micro to nano, giving exceptional properties. Nano sheets, which are crystalline when embedded in an amorphous matrix, give to wide range of designs suitable for the production of fibers and advanced composites for aerospace applications, military applications, automobile, and nuclear applications.

Multicomponent materials like composites, alloys automobile offers multi-functionality. For example, aluminum is a soft material and corrosion-resistant, but its mechanical strength is very weak, and so are its wear properties. On the other hand, silicon carbide has high mechanical strength and high wear resistance [17]. So, these materials are combined to give a multi-functional material, which is wear-resistant, as well as having good mechanical properties. This brick-and-mortar structure can be found in nacre, which is composed of aragonite platelets. This configuration is highly fatigue and fracture-resistant. Many sea organisms, such as corals and sea urchins, can create mineralized structures. First, they extract calcium and carbonate ions from seawater, and then they convert them into solid calcium carbonate, and using proteins and polysaccharides, they create a skeleton of desired shape, size, and even orientation. These techniques are done at room temperature. This inspired researchers to build ceramics and other mineralized materials at room temperature, which would otherwise use traditional high-temperature sintering processes.

Dynamic and stimulus-responsive materials are plentiful. Plants can exhibit mobility and move their leaves or petals and track sunlight (sunflower plant) for better photosynthesis. Again, sunflower is a composite material consisting of thousands of tiny seeds called florets. Pines respond dynamically to changes in humidity in nature. When the humidity is high or a wet atmosphere is sensed by pines, it closes its cones so that seeds will not fall out. On the other hand, if the atmosphere is dry, then the cone opens wide so that the seeds spread apart. A cone consists of two layers, inner and outer. The inner layer can absorb more moisture and also swell more, and the outer layers absorb less moisture and swell less. When there is high humidity in nature, the inner layer expands more than the outer layer, which causes bending of pines and closing. Opposite things happen when there is less humidity in nature [18]. Engineers used this principle and developed hydrogels, which can be used in sensors and actuators for self-regulation, such as shape memory polymers for biomedical and robotics surgery. Smart textiles and self-opening ventilation systems were developed based on these principles. Nature gave tremendous inspiration by creating mussels. Mussels can produce a protein-based amino acid strong glue, which is safe even underwater. These bonds are a complex combination of hydrogen bonds, metal coordination bonds, and covalent cross-links. Engineers used these principles to produce underwater construction materials, medical adhesives, coatings for marine applications, and many more.

There are many self-cleaning materials that nature has offered to mankind to learn technically and spiritually. A lotus leaf is a highly hydrophobic material. When the water falls on a lotus leaf, the water rolls off, simply carrying away the dust on the leaf, leaving the leaf clean. No external force is needed to remove the dirt or dust from the leaf. Paint manufacturers derived inspiration from these principles and developed self-cleaning paints [19]. Lotus leaf contains roughness on a micro scale and wax crystals on a nano scale. Wax coating is highly water repellent, and in total dual structure makes a leaf superhydrophobic. This principle is also used in self-cleaning solar panels, textiles, medical devices, eyeglasses, waterproof jackets, and mobile phones. Any product or device functionality depends on a combination of good design principles, good materials, and good manufacturing processes. Many fabrication and manufacturing processes, such as additive manufacturing processes, composite fabrications, are inspired by biological fabrication processes available in nature, such as multiple tissues of organisms [20]. Nature does not cut materials but builds material, which is the building block principle in additive manufacturing. Honeycombs, birds' nests, spider webs, enamels, and bamboo are just a few examples. Self-assembly is nature's elegant and most useful principle, which has provided huge inspiration to human beings. Proteins naturally fold into 3D shapes depending upon the sequence of amino acids, natural alignment of lipid bilayers, helix of DNA, and are simple examples that resulted in the most sophisticated engineering products, such as drug delivery systems, sensors, nano electronics, photonic crystals, and smart materials.

Apart from additive manufacturing, electrospinning is yet another bio-inspired manufacturing process used to produce bioactive compounds that can be found in bio-integrated medical devices. Inspiration is obtained from the extracellular matrix found in tissues. The advantages of electrospinning are high surface area, good resolution, controllable porosity, and density of fibers, which can be used in several tissue engineering applications, as well as membrane filtration techniques commonly used by chemical engineering. Now, flexible electronics are being researched by a large number of researchers for wearable devices that house with necessary electronics. Skin is the most bio-inspired material for flexible electricity. Its elasticity, dexterity, plasticity, impact, and fatigue resistance can be engineered to get implantable devices, medical polymers, electronic skins, health monitors, and chip-on-board devices, which provide bio-flexibility.

The industry needs more sustainable and eco-friendly green products, such as biodegradable biopolymers. Researchers are interested in developing a material that can degrade under specific environmental conditions, be recyclable, reusable, and multi-functional, so that waste is minimized. Polylactic acid and chitosan, which are available in nature, are biomaterials that are now used in developing sustainable packing materials to replace harmful plastics. Bio composites are used in automobiles as their weight are lighter and stiffer. The textile industry mimics sharkskin to produce underwater suits that have low drag forces. Many paints are developed to protect from microbes and are water-repellent based on bioinspired materials available in nature.

1.6 Bio-Inspired Devices in Engineering Systems

Effective bio-inspired devices can be realized when biologists, material science experts, mechanical engineers, electronics engineers, and doctors work together, as it is truly a multidisciplinary system. The teamwork of all disciplines can design devices that are not only efficient but also sustainable, resilient, contextually adaptive, multi-material, and multi-functional by drawing wisdom from nature and by using available state-of-the-art technology. Robotics engineers can design simple robot-like pick and place to complex robot-like terrain run robots by mimicking insects, animals, and even cellular organisms. Locomotion, sensing, actuation, workspace optimization, route/path planning, self-correction, and self-healing modules are introduced in robotics to achieve breakthroughs in these technologies. Some robots can climb vertical walls; these designs are based on the gecko's foot. Deeper insight into the adhesion processes, van der Waals forces, and action of setae in the microscopic range for gripping and locomoting in the desired direction led to these kinds of military and construction robots [21]. These types of robots can be efficiently used in construction industries for lifting, transportation, plastering, plumbing, inspection, mixing, search and rescue operations, and maintenance of tall buildings, like cleaning, painting, and repairing.

Highly flexible, pliable, and deformable organisms are worms, which started the octopus-inspired development of soft robots. They use flexible structures, materials, fluid-operated actuators coupled with response stimuli sensors to move and manipulate objects, which are in the available workspace. Since they are highly flexible, they can travel where AGV and rigid robots cannot travel, like narrow tunnels, nuclear power plants, underwater, invasive surgery, and prosthetics. They are adaptable and dynamic and can be operated and observed in real-time, as in surgery using flexible micro-robotic systems [22]. In aerospace engineering, safety, efficiency, maneuverability, surveillance, mapping, monitoring, sensing, and actuation are extremely important. Birds are the main inspiration for aerospace development, be it a sensing capability of eagles/bats, maneuverability capability of hummingbirds, safety like huddling of penguins, efficiency like albatrosses, surveillance like eagles, hawks, falcons, mapping like pigeons, monitoring like vultures. Millions of lessons can be learned by studying birds.

Swift and Falcon possess excellent variable wing geometry to reduce drag and increase lift, which led to the invention of the Morphing Wing Aircraft. During landing, birds separate their feathers to reduce speed by manipulating hydrodynamically with a stable landing. Albatrosses and vultures can cleverly adjust their structure to wind currents to maximize efficiency and minimize the energy expenditure [23]. Hummingbirds and swallows have excellent maneuverability even in a very small space. Eagles' acute eyesight or owls' spatial hearing promoted the design of UAVs, which are very useful in military, agricultural mapping, crop management, spraying, disaster monitoring, and many more. Flexible piezoelectric sensors are fabricated by taking inspiration from bird feathers, which can sense and detect flow separation and adjust curvature control surfaces optimally. Geese fly in V-formations, which is hydrodynamically efficient since such V-formations can give minimum drag and thus flight energy can be saved. Hollow structures are stronger and stiffer than solid structures, and birds' bones have internal structures called trabeculae, which give weight reduction and a high strength and stiffness ratio, which are particularly useful in aerospace applications. Termite mounds provide a great inspiration to architects and civil engineers. Mounds are very efficient tall structures with a unique chimney and side ventilation. Many thermal management systems used by termites are mimicked in buildings like the Easter Gate Center in Harare. Termite mounds are natural marvels of science, engineering, and technology combined. Mounds provide a stable internal temperature irrespective of outside temperature through fungus formation and gas exchange [24]. Passive ventilation systems of termites consist of porous walls and stacking so that cold air comes in, and hot air goes out due to the density difference. Termites use an excellent combination of proper soil, saliva, and plant matter to produce thermal inertia composite materials to maintain a stable temperature. It is no surprise that termite mounds use 95% less energy compared to a conventional building, which has similar facilities.

Medical needles are developed by observing mosquito bites with minimal pain and minimal tissue damage by plastically deforming the skin slightly during insertion. Micro needle patches are used to deliver drugs, vaccines, and insulin to the human body in a minimally invasive fashion. Prosthetic limbs are developed by observing musculoskeletal systems, which have excellent joint actuation mechanisms and tendon units, which work with muscles in unison. Nature has had a profound impact on the development of advanced sensor technology, compound eyes of insects, and mirror-like structures, inspiring the development of projector lenses. The concept of compound eyes is effectively used in liquid detection systems, photoreceptor cells, which can be used for partially restoring vision and reducing the risks of degenerative eye diseases [25]. The skin of birds, humans, and other organisms inspired many sensor inventions, such as pressure, temperature, humidity sensors, and health monitoring sensors.

There is a severe energy and water crisis due to depletion of natural resources and man-made calamities like wars, terror attacks. Energy production and reduction of wastage are important. Plants produce energy by using the light of the sun and convert it. These efficient energy conversion concepts are used in solar energy devices, and they fuel cells. Electricity can be generated from the waste. Nature's closed-loop philosophy, if adopted, would make energy conversion easy and efficient. In high-speed trains like bullet trains, aerodynamics plays a very important role in reducing fuel consumption. The fuel consumption can be reduced when there is less resistance to the movement of the train. Japanese engineers mimicked the kingfisher's beak to design the front horn of the bullet train, and it is not efficient train but also produces low noise during its travel in tunnels. In ships, special textures resembling shark skin are produced, which can reduce the drag force and increase the efficiency [26]. These two examples show how bio-inspired design changes transportation engineering; there are many such examples in automobiles, as well.

Computer science and information technology also use bioinspired design. The structure of the brain and the flow of neurons in human systems gave rise to AI techniques such as neural networks. Many algorithms and distributed systems are inspired by ants, bees, and birds. Many concepts, like mutation and crossover, are taken from cellular organisms in optimization techniques like genetic algorithms. The living style of many organisms is used in inventory control, machine scheduling, in operational research problems. There are many machine learning algorithms that use collective behavior of organisms to adapt, train, and gain expertise with the use of data collected either through experimentation or by artificial intelligence built into the algorithms themselves. Bio-inspired devices not only allow us to mimic form but also many things like function, materials, structure, design, stability, feasibility, adaptability, sustainability, intercommunication, imaging, ecological harmony, discipline, constraints, scalability, contextuality, guidance, survival, and many more [27].

1.7 Bio-Inspired Devices in Medicine and Healthcare

The field of medicine and healthcare uses many instruments, devices, and apparatus, which are designed, fabricated using bio-inspired design and systems. There are three important considerations in healthcare and medicine. They are healing, preventing the spread of diseases, and providing rehabilitation. Nature provides genuine mechanisms for adaptation, reuse, repair, and survival models, which can be used to produce many surgical instruments and diagnostic devices, drug delivery, dental and bone implants, and wearable and flexible electronics. Many innovations in healthcare are bioinspired. The human body itself is a source for many innovations like pumps, neural networks, and distribution systems, sensors, locomotion, and transpiration. Today, the advancement in microscopy has made it possible to observe small organisms and insects in greater detail. A close observation of a mosquito's proboscis reveals that the mosquito has a needle-like structure, which is a functionally graded microstructure that can vibrate, and the needle penetrates without much pain and resistance to the person. This inspired the development of micro needles, which consist of microstructures ranging from 100 to 1000 microns in serrations. Silicon, stainless steel, biomedical polymers, or sugars are used to manufacture these micro needles. There are dissolvable microneedle patches, which can be used to administer many vaccines without using a syringe. This is safe, avoids contamination and infection due to fear of reuse of syringe, solid and coated microneedles are under research, which can be used to deliver drugs promptly.

Many microneedles are made from the drug to be delivered itself. Bio-additive printing, micro molding, and advanced lithography techniques can be used to make such microneedles. Drugs are delivered through the micro needle, which injects the drug between the EPI and upper dermis layers by creating sophisticated micro channels for the easy flow of drugs in the body, and pharmaceutical contents easily diffuse into the body through these microchannels with quite high precision and accurate dosage. Hydrogel micro needles cause the skin to swell upon injection, and hollow needles are fabricated with inspiration taken from blood sucking organisms. Shape memory alloys, which are response sensitive to load, pressure, and temperature, are also used for effective drug delivery. Drug delivery systems are used in insulin injections for diabetes, chemotherapies for cancer. There are chip-on-board systems developed, which are bio-inspired to check lactate content, blood pressure, and blood sugar. Micro needs are designed to combine many functions so that cost is reduced, and the highest therapeutic efficacies are achieved through patient comfort [28]. Transformative medicines and healthcare are possible by mimicking biological functions, mechanisms, working, and behavior.

The biting mechanism of animals, insects, organisms, and humans provided a greater insight into the design of scalpers and biopsy tools. Structures of scalpers are mimicked by observing the teeth of sharp and mandibles of beetles. They have ultra-sharp edges, which can cause minimal pain to the patients. Robotic surgery using micro robots mimics cephalopod limbs, which are highly flexible and dextral, allowing precise laparoscopy and microsurgery, which increases the survival rate of the patient. Prosthetics are bio-inspired by human anatomy and biomechanics of the human leg, knee, and ankle. The mechanism of the knee joint, ankle, and hip provided bio-inspired implants that are not only safe but also biocompatible. Cheetahs are gifted with highly elastic energy-absorbing spring-like tendons and are mimicked by companies to produce flex feet, which allows amputees to run at high speed, especially in competitions meant for gifted people and the mentally challenged. These flexible feet use composites, which use carbon fibers to give good strength and also a good energy absorption and storage structure, like a spring, which can give remarkable speed, as well as running efficiency. Researchers invented a healthcare medical device that can have nerve and muscle interfaces found in organisms like octopus and spiders.

Plant structures, such as lines, vines, and roots, inspired the development of advanced implantable pacemakers. These are designed in such a way that they get dissolved in the human body over time. It is time-tested and safe, which provides support to blood vessels and improves the life span of heart patients. Anti-bacterial coating on the implants is prepared by taking inspiration from the self-cleaning properties of lotus leaves, which reduces the infection during and after the surgery. Over the years, endoscopy has evolved due to advancements in semiconductor devices. Wide-range cameras are now used in endoscopy, which are developed by observation of the compound eye structure found in insects and other micro and macro-organisms. This provides a good observation of the internal organs of humans and animals, and thus early detection of diseases, infections is possible, which can save lives. Fishes and spiders are gifted with vibration-sensing abilities, which are mimicked in biosensors used for health monitoring devices in disease management, especially in remote areas.

Regenerative medicines save many lives every year. These are nothing but bio-inspired medicines. Scaffold design and modulations of cells are mimicked by studying the cell behavior of sea cucumbers and starfish. The scaffold must be stiff and soft. Both these contradictory properties can be obtained by using advanced manufacturing technologies. Many innovative human parts, like skin, cartilage, and valves, are made by using materials similar to spider silk, plant-derived proteins through bio-printing processes. These designs are also capable of nutrient management, growth of cells, and adherence to parent bones or parts of the body. Artificial respiration systems are developed by mimicking fish gills. The gliding and sliding mechanism used by birds is useful in the design of stents, which can expand due to external response or stimuli. Nanoparticles can travel fast in the bloodstream, and quick drug delivery is possible. Modern technologies are available to make the drug into nanoparticles, and they can be introduced into the body to reach specific parts. The navigation procedure of nanoparticles is inspired by the movement of spread of bacteria and viruses inside the bodies of animals and humans. The movement of particles can be controlled to reach specific areas supported by external stimuli, such as a magnetic field, to cure tumors, cancers, and infections. Broader imaging techniques for medical observation are developed by mimicking frog eye lenses, and many stimulus-response materials like shape memory alloys are used for expandable devices that are useful in medical devices.

1.8 Bio-Inspired Devices in Environmental and Agricultural Applications

Sustainable Development Goals give importance to affordable and clean energy. Environmental pollution can be controlled by monitoring and innovating in decarbonization and developing sustainable agriculture technologies. It is through plants and forests; environmental balance can be achieved, and to a certain extent, clean water can be obtained by conservation principles available to us in nature. Burning global issues are water scarcity, degradation of soil, excessive use of pesticides in agriculture, dramatic climate change due to heavy industrialization, and an imbalance in the ecosystem on Earth. Superior sensory systems available in many organisms and animals, like seals and rats, are effectively mimicked to develop a new class of sensors, which can detect the vibrations and water movements [29]. Sensors can detect water quality and register the marine life quality based on water contamination or purity.

Surveillance is not only important in the military but also important in agricultural crop monitoring, land monitoring, sea monitoring, water monitoring, humidity monitoring, and many other applications. Surveillance cameras are needed to cover a wide range of land or space. Wide-angle cameras are now available, which are bioinspired by frogs' eyes, dragonflies, bees, and the compound eyes of minute organisms. The secrets revealed during the microscopic study of organisms using high-resolution SEM and TEM provide a wealth of information about the lens arrangement and optimum structures needed to house them. Crop monitoring, spraying, and soil testing using cameras mounted on drones are done effectively. Early forest fires can be detected using fire sensors.

Air filters and water filters are essential to provide clean air and water, which are basic for human survival. Filters are developed by taking bio-inspiration from fish gills, which can even filter out particles of nano size and volatile materials. Many solar panels use self-cleaning agents, which are superhydrophobic, inspired by lotus leaves, which use minimum water. Beetle provides an excellent water harvesting mechanism. There are two major mechanisms: one is condensation, and the other is collection. For condensation, beetles have hydrophobic surfaces, and for collection, beetles have hydrophilic surfaces. Not only are these surfaces hydrophobic, but they are also graded in the micro to nano scale, which makes it a more interesting water harvesting mechanism for humid environments. Engineers have used this mechanism to capture water from fog through a fog harvesting mechanism similar to beetles. This type of water harvesting mechanism produces clean water, and it is environmentally friendly and sustainable. The signal processing of some insects is now used in agricultural pest management. They mainly use pheromones made of synthetic materials and visual aids. Many scare devices are designed to prevent crop destruction by birds, organisms, and wildlife. Soil conservation and nutrient preservation are challenging for farmers. The distribution of nutrients is important to get uniform crops and also to get uniform absorption of moisture and water by plants. Many precision agriculture technologies are bio-inspired. Bee flight and flower interaction inspired many developments in agricultural equipment, such as pollination drones, robotic pollinators.

Plants use their leaves and convert solar energy to chemical energy by a process known as Photosynthesis. Now researchers have successfully designed and developed artificial leaves which is like plant leaves, which can convert solar energy to chemical energy. They are further used to get water, carbon dioxide, and other organic compounds. It is fascinating to observe how bacteria established their self-defense system against a certain class of viruses. Similar technology is now used to modify the plant genomes.

1.9 Bio-Inspired Robotics and Autonomous Systems

Robots are not only used in industries but also in many medical operations. Their size varies from very large to micro. Micro-robots are commonly used in invasive surgeries. Biological organisms, such as fish, arthropods, quadrupeds, ants, cockroaches, dogs, birds, bats, insects, and geckos, inspire design, manufacture, control, programming, and locomotion. Sensory integration and machine communication with a robot is mimicked using the principles of locomotion of animals and organisms. One inspiration is from insects. Robots can now navigate in complex environments, move like dogs and donkeys with kinematics similar to dogs' legs. They are smarter, equipped with a real-time control system that offers exceptionally good features, including obstruction management, voice modulation, touch sensing, cooperative behaviors, good dexterity, resilience, adaptability, and the ability to sense neuromechanical stimuli.

Engineers effectively mimicked the locomotion of robots by simulating walking gaits of dogs, cockroaches, centipedes, crawling creatures, snakes, snails, ants, termites, and many more. Sophisticated software is available along with high-speed cameras to capture the gait cycles of the fastest running animals, like cheetahs, and they can be easily modified in the lab to get the most suitable locomotion for robots. Many robots can be used for bomb disposal and toxic material handling in very difficult terrain with canine gait control and morphological dynamics. Many balancing algorithms are developed for controlling the stability of the robots using advanced control strategies, using machine vision and sensors coupled with the traditional control stability strategies and real-time actuations. Dragonflies are well known for their excellent maneuverability in small and complex spaces. This principle is used in developing micro flying flapping wing drones [30]. The flying dynamics of bees are studied in detail, and actuator and sensors for military applications are available, which weigh less than a few grams.

Ant colonies, bees, and birds inspire many optimum algorithms. Their network is complex and involves multi-functional objectives involving many constraints. These principles are applied in engineering by building swarm robots, multi-agent robots, which can now be used for mapping of very large floor areas and are helpful in natural calamities to find people, animals, so that food and water can be supplied during flood situations. Underwater robots are developed by mimicking jellyfish and its tentacles. Flexible spines and actuators in tuna are adopted in underwater robots. Micro robots, which resemble microorganisms, are now used in chemical pipeline inspections and marine life tracking. Octopus arms, elephant trunks, and starfish legs inspired engineers to develop soft robotics. They are widely used in surgery and are capable of good grasping, bending, and holding. They are adoptive, like squid tentacles, and can pick up irregular objects and fragile objects like glass. The principle of tactile sensors present on gecko feet and spider hairs inspired engineers to develop vertical wall-mounting robots, which can be used for tall structures for monitoring, cleaning, and inspecting.

1.10 Conclusion, Future Trends, and Challenges

Bio-inspired designs are here to stay as they promise the most advanced devices and instruments through a combination of natural intelligence, engineering, medicine, biology, and many more branches of science, engineering, and technology. Digital biomimicry is one of the emerging fields in bio-inspired design. With an advancement in digital manufacturing, inspection, and digital characterization, it is possible to get more advanced generative design through advanced computational models mimicking the organisms and complex biological processes. Many advances in materials innovations have led to self-repair, flight mechanics, self-healing, self-cleaning materials, and self-transforming, which will change the smart textiles, devices, and structures into a more sustainable, adaptable, and energy-efficient state. Response stimuli materials will change the field of sensing and actuation. Shape memory alloys, biocomposites, and biopolymers for medical applications bring innovation in devices and machines. Soft robots' technology, bio bots, and bio hybrid systems will bring innovation in microsurgery, drug delivery, and climate sensing.

The field of regenerative design will change urban planning, building structures, safety, and traffic control. Miniaturization and nanotechnology will further develop and cross the boundary of bio-inspiration. The challenge is to transform design ideas into a manufacturable product. There is a long way to go in this process, and humans will succeed through bio-inspiration.

References

1. Vincent, J.F.V., Bogatyreva, O.A., Bogatyrev, N.R., *et al.*, Biomimetics: its practice and theory. *J. R. Soc. Interface*, 3, 471–482, 2006.
2. Bar-Cohen, Y., *Biomimetics: Nature-Based Innovation*, CRC Press, Boca Raton, 2011.

3. Bhushan, B., Biomimetics: lessons from nature—an overview. *Philos. Trans. R. Soc. A*, 367, 1445–1486, 2009.
4. Gebeshuber, I.C. and Drack, M., An attempt to reveal synergies between biology and engineering mechanics. *Proc. Inst. Mech. Eng. C*, 222, 1281–1287, 2008.
5. Fratzl, P. and Weinkamer, R., Nature's hierarchical materials. *Prog. Mater Sci.*, 52, 1263–1334, 2007.
6. Chen, H., Müller, M.B., Gilmore, K.J., *et al.*, Mechanically strong, electrically conductive, and biocompatible graphene paper. *Adv. Mater.*, 20, 3557–3561, 2008.
7. Kim, S., Laschi, C., Trimmer, B., Soft robotics: a bioinspired evolution in robotics. *Trends Biotechnol.*, 31, 287–294, 2013.
8. Santos, D., Nanomechanics and biomimetic surfaces: towards micro- and nanorobotic applications. *Microsyst. Technol.*, 17, 831–846, 2011.
9. Del Dottore, E., Mondini, A., Mazzolai, B., Growing robots: bioinspired systems for a sustainable future. *Annu. Rev. Control Robot. Auton. Syst.*, 3, 301–330, 2020.
10. Mazzolai, B., Laschi, C., Cianchetti, M., *et al.*, Soft-robotics technologies to address shortcomings in today's minimally invasive surgery: the STIFF-FLOP approach. *Soft Robot.*, 1, 122–131, 2014.
11. Li, J., Esteban-Fernández de Ávila, B., Gao, W., *et al.*, Micro/nanorobots for biomedicine: delivery, surgery, sensing, and detoxification. *Sci. Robot.*, 2, eaam6431, 2017.
12. Seok, S., Onal, C.D., Cho, K.-J., *et al.*, Meshworm: a peristaltic soft robot with antagonistic nickel titanium coil actuators. *IEEE/ASME Trans. Mechatron.*, 18, 1485–1497, 2013.
13. Mahdavi, A., Ferreira, L., Sundback, C., *et al.*, A biodegradable and biocompatible gecko-inspired tissue adhesive. *Proc. Natl. Acad. Sci. U.S.A.*, 105, 2307–2312, 2008.
14. Roxhed, N., Samel, B., Nordquist, L., *et al.*, Painless drug delivery through microneedle-based transdermal patches featuring active infusion. *IEEE Trans. Biomed. Eng.*, 55, 1063–1071, 2008.
15. Mishra, R. and Vashisth, S., Bioinspired microneedle design for painless drug delivery: a review. *J. Micromech. Microeng.*, 28, 053001, 2018.
16. Helms, J.A. and Kaplan, D.L., Bioinspired design and biomaterials for regenerative medicine. *Wiley Interdiscip. Rev. Dev. Biol.*, 2, 97–112, 2013.

17. Trivedi, D., Rahn, C.D., Kier, W.M., *et al.*, Soft robotics: biological inspiration, state of the art, and future research. *Appl. Bionics Biomech.*, 5, 99–117, 2008.
18. Vogel, S., *Cats' Paws and Catapults: Mechanical Worlds of Nature and People*, W.W. Norton & Co., New York, 2000.
19. Lakhtakia, A. and Martín-Palma, R.J., *Engineered Biomimicry*, Elsevier, Waltham, 2013.
20. Festo, Bionic Learning Network, Bionic Soft Arm and Smart Bird projects, 2022. URL: https://www.festo.com/group/en/cms/10156.htm.
21. Zhang, Y., Zhang, L., Yang, X., *et al.*, Bioinspired artificial spider silk: combining strength and elasticity. *Adv. Mater.*, 30, 1703658, 2018.
22. Grassé, P.P., Termitidae and the termitophile relationships. *Ann. Sci. Nat. Zool.*, 11, 209–270, 1959.
23. Lee, S. and Lee, D., Bioinspired strategies for effective and sustainable water collection. *Small*, 17, 2105694, 2021.
24. Wang, X., Liu, X., Duan, Y., *et al.*, Bioinspired surfaces with switchable wettability. *Chem. Rev.*, 115, 8230–8293, 2015.
25. Bartels, T. and Morris, J.M., Case study: Morphing aircraft wings inspired by birds. *J. Aircr.*, 51, 1066–1073, 2014.
26. Torres, F.G., Troncoso, O.P., Diaz, D.A., *et al.*, Gecko-inspired dry adhesive: functional fabrication and applications. *Adv. Funct. Mater.*, 24, 588–596, 2014.
27. Park, J., Kim, J., Lee, D., *et al.*, Microfluidic technologies for neuronal cell culture: from bio inspiration to applications. *Lab Chip*, 13, 509–521, 2013.
28. Helbig, R., Nickerl, J., Neinhuis, C., *et al.*, Smart skin patterns protect springtails. *Science*, 339, 932–935, 2011.
29. Genzer, J. and Marmur, A., Biological and synthetic self-cleaning surfaces. *MRS Bull.*, 33, 742–746, 2008.
30. Huang, J., Wang, X., Wang, Z., *et al.*, Bioinspired photonic structures with tunable structural colors. *Nat. Commun.*, 6, 6979, 2015.

2

Bio-Inspired Manufacturing and Fabrication Processes

Abstract

Nature is a great inspiration for innovation, product development, sustainability, and exceptionally superior technologies. Many researchers mimic bio-inspired designs, like the superhydrophilicity of beetles, the structure of spider webs, the material aspect of nacre, the colonization of ants, the building structure and ventilation of mounds by terminates, and many more. Proper manufacturing and fabrication technologies are needed to realize these bio-inspired designs. This chapter explains how bio-inspired design is translated into a product or device through advanced fabrication techniques. A comparison between conventional processes and bio-inspired fabrication is discussed with real-world applications inspired by gecko feet, lotus leaves, termites, shark skin, spider webs, ants, bees, elephants, etc. Types of manufacturing depend on the function and material. This chapter also discusses material aspects and available advanced fabrication technologies, like additive manufacturing, soft lithography, photolithography, bio printing, electro-spinning, laser-based processes, protein-based processes, with detailed comparisons between material efficiency, suitability, eco-friendliness, adaptability, energy efficiency, and customization. A superior and smart integration of biology with various branches of engineering gives innovative technologies, products, devices, instruments, and machinery for the future of humankind.

***Keywords*:** Bio-inspired manufacturing, additive manufacturing, sustainable fabrication, biomimicry, functionally graded materials, soft lithography

Jagadeesha T. and Sandip Kunar. Bio-Inspired Manufacturing Systems, (31–56) © 2026 Scrivener Publishing LLC

2.1 Introduction

Nature is always a source of inspiration for human existence, evolution, and innovation. Human beings initially copied nature by pure observation for survival. He observed how animals hunt and make their food, and he used stone weapons to kill animals for his food. He observed how animals survive during the rainy seasons by living in caves, so he copied nature and started to live in caves. After observing the bird's nest, he started to build his own house using straw and mud. Initially, human beings just mimicked nature for their survival. As time progressed, he evolved with nature and started his innovation. He observed birds' superior aerodynamics and built aircraft. Inspired by gecko feet, advanced glue and vertical wall-climbing robots were invented. Bioinspiration is a multi-disciplinary domain that uses biology, medicine, and all types of engineering to convert scientific principles observed in nature into technological innovation, devices, or products. This chapter introduces the manufacturing and fabrication processes available to realize bio-inspired design, processes, or machines. Fabrication processes are needed to convert ideas into devices, so that the product is functional, adaptable, contextual, structurally and performance-wise efficient, and sustainable with minimal environmental impact [1]. Any bio-inspired fabrication process must capitalize on nature's wisdom because they are time-tested and energy-efficient with minimal use of parts. Even though nature has evolved over billions of years, it still has many wonders to reveal. For example, with advancements in microscopy, SEM, and TEM, we can observe many micro-, nano-graded structures on beetles, the structure of lotus leaves, and the vibratory control mechanism of spiders, the strength and modular design of the spider web, and more. As we progress in science and technology, we will discover more, appreciate nature more, and create more things by bioinspiration.

Bio-inspired manufacturing processes differ from traditional processes; most of the processes used for bio-inspired fabrication are advanced techniques, and they do not use many conventional processes due to the poor efficiency of traditional processes. A comparison is given in Table 2.1.

Three concepts are combined to make bio-inspired products. One of the classical approaches is biomimicry, in which the structure, motion, and function of animals, plants, and organisms are imitated. The second approach is called bio-utilization, wherein biomaterials are directly used in the fabrication and manufacture of bio-inspired devices or products. The third approach is bio-assisted manufacturing, in which bacteria and enzymes are part of the fabrication processes. A deep understanding of the principles beyond the behavior

Table 2.1 Comparison between bio-inspired and conventional fabrication methods.

Sl. no.	Fabrication criteria	Bio-inspired fabrication	Conventional fabrication
1	Energy consumption	Most of the fabrication processes are low-temperature, low-strain processes, and energy consumption is low	In subtractive processes, material is removed through high-strain rates and high-temperature processes, and energy consumption is high.
2	Usage of material	Minimum as in additive manufacturing	The maximum amount of material is wasted in the form of chips in subtractive processes.
3	Degree of adoptability	Dynamic in nature, using real-time data	Static in nature.
4	Eco friendliness	Eco-friendly	High energy-consuming process, so not eco-friendly.
5	Degree of complexity	A highly complex component can be fabricated	Difficult to make complex parts and expensive.

of animals and organisms is necessary for fabrication, and it is not a simple copying in raw form. Materials, environment, resources, energy requirements, adaptability, manufacturability, cost, and availability must be considered along with other design constraints to realize the component that is bio-inspired. Pros and cons of conventional and non-conventional methods must be examined before choosing the fabrication processes [2]. Conventional methods use

Table 2.2 Natural principles and products developed.

Sl. no.	Natural principle	Source/organism/ animal inspired	Product/devices/ materials developed
1	Adhesion and vertical climbing	Gecko foot structure and gaiting	Advanced adhesives and vertical wall climbing robots
2	Graded material structure	Shark skin micro- to nanostructures	Swimsuits and super-hydrophobic coatings for marine applications
3	Super hydrophobicity – lotus effect	Structures available in lotus leaves and beetles	Self-cleaning textiles and water harvesting in deserts
4	Hollow structs and graded	Bones of birds	Composite materials, aerospace structures
5	Complex lightweight structures	Honeybees	Tall buildings and aerospace applications
6	Temperature and ventilation	Termite mounds	Tall building with self-ventilation and temperature control
7	Lightweight fabrics	Spider silk	Synthetic fibers are stronger than steel
8	Photosynthesis	Leaves of plants	Artificial leaves, solar panels that turn optimally to sunlight like the leaves of plants

lots of resources, and the waste produced, such as in the form of chips, is high. On the other hand, bio-inspired processes have less carbon footprint and use non-renewable materials and processes. Table 2.2 shows some of the natural principles and sources in nature and how engineers use them in technology and product development.

Bio fabrication involves effort from biologists, chemists, material scientists, engineers, and doctors. For example, water not clinging to lotus leaves was observed by biologists, and superhydrophobicity was measured by material scientists, chemists, and engineers who made artificial paints that are self-cleaning, as shown in Figure 2.1. Now, this natural principle is applied in self-cleaning solar panel coatings and the textile industries. This shows how a simple phenomenon can be translated into to useful product and also shows multidisciplinary attention to develop those devices or products which are energy efficient and adaptable [3]. Self-cleaning paints avoid repainting to a large extent and reduce environmental impact if not completely.

Figure 2.1 Self-cleaning effect or lotus effect.

2.2 Biological Principles Used in Bio-Inspired Manufacturing

To make or manufacture any product, it is important to understand the basic principles, phenomena, behavior, and mechanisms of the system beforehand. Better devices or products can be manufactured by maximizing resources by taking care of process, system variability, and market demand. Nature adopted multi-material and multi-functional structures and uses minimum parts and resources. But, manufacturing multi-material and multi-functional in one system may not be possible, and two or more materials and processes are needed to make the same as in the case of functionally graded materials. Natural materials in plants, animals, and organisms have a multiscale arrangement [4]. For example, in beetles, their exoskeleton structure has both micro- and nanostructures, one for capturing dews from fog and the other for water collection to thrive in deserts, as shown in Figure 2.2.

Another example of a multistructure is the human bone, which is a composite material made of structured material that consists of

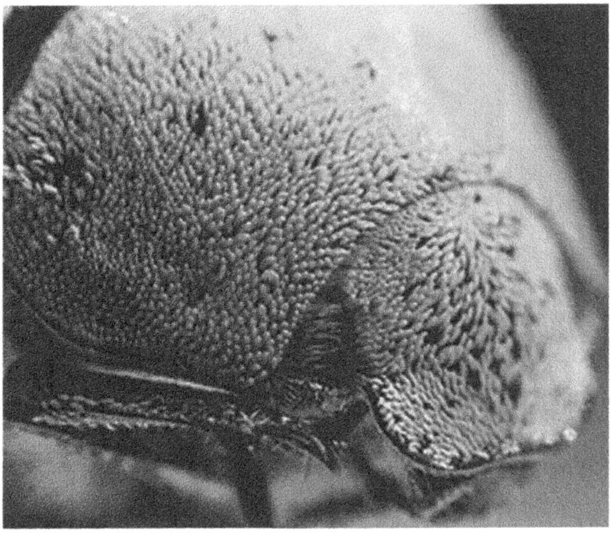

Figure 2.2 Graded structures in beetles.

collagen and hydroxyapatite. Many bones that inspired multi-layer composite materials are used in aerospace applications. Graded materials and meta materials are also developed by studying the bone structure microscopically and macroscopically. The skin of a shark provides an example of a multi-functional material. The curvature reduces the drag, and the skin also protects the shark by preventing the growth of microbes; it also has superior biosensing abilities. This led to the growth of medical implants with a specific coating that avoids the growth of bacteria and microbes. Principles of the formation of helices of DNA and protein folding are widely used in micro- and nano-fabrication. Self-aligned materials are used in wafer fabrication in the semiconductor industry. The skin of animals and humans is a good biosensor and has a good responsive stimulus, like sweat and goosebumps, which occur due to temperature and pressure changes within the body or environment. Many smart materials shape memory alloys and are developed using this

Table 2.3 A summary of biological principles and products developed.

Sl. no.	Principle adopted	Inspiration	Product/material/device
1	Structuring - gradient	Multiscale organization, such as a beetle or nacre	Composite material, functional graded material, lattices
2	Multi-functional	Single structure with multifunction	Intelligent materials, coatings
3	Self-assembly	Unique organization systems	Nanomanufacturing, functional
4	Adaptability	Stimuli-changing materials	Smart textiles, shape memory alloy
5	Material efficiency	Hollow structures, such as bones	Green, lightweight materials.
6	Topological optimization	Principle of minimum material usage	Lightweight material for aerospace and automobiles

principle [5]. Spiders use only water-soluble proteins without using any temperature or high pressure, yet produce a silk stronger than steel. This led to the invention of biopolymers, green synthesis, and bio-inspired concrete. Table 2.3 gives a summary of a few biological principles and products developed based on these principles.

2.3 Bio-Inspired Fabrication Strategies and Technologies

Any design must be manufacturable. Design for manufacturing and assembly are important considerations during the design process, whether conventional or bio-inspired. Manufacturing processes must be cost-effective, technologically advanced, have a shorter lead time, be of good quality, be eco-friendly, sustainable, and be able to produce in large quantities to take advantage of economies of scale and cost [6]. The following manufacturing processes are usually used in bio-inspired products, instruments, devices, materials, and machinery.

1. Additive manufacturing or 3D printing
2. Functionally graded materials manufacturing
3. Soft lithography
4. Direct self-assembly methods
5. Bio printing
6. Microfluidic fabrication
7. Bio-mineralization
8. Origami-inspired design
9. Bioelectronic interfaces

2.3.1 Additive Manufacturing

It is also famously known as 3D printing because it can print 3D objects by adding materials layer by layer. In this process, a material is added instead of being removed as in subtractive processes; therefore, a loss of material is minimal. The first part to be manufactured is created using computer-aided software, and then CAD data

Table 2.4 Comparison of traditional and additive manufacturing processes.

Sl. no.	Parameters	Conventional method	Additive manufacturing	Usefulness of bio-inspired design
1	Degree of complexity	Complex parts cannot be made and limited by die shapes	Complex parts can be manufactured easily	Most biological structures are complex in shape
2	Material usage	Subtractive processes have high wastage of material	Wastage is minimal	Eco-friendly
3	Tailor made	Customization is difficult	A single part can be produced	Personalized bio-inspired design is possible
4	Multi-material usage	It is difficult to integrate two or more materials	It is easy to print two or more functionally graded materials	Multi-component and functionally graded materials offer good advantages
5	Design processes	Process and modification are difficult	Easy, as only CAD models are changed and not tools and dies	Short lead time and product/process variation are possible to meet the ever-changing demands of the market

is digitally sliced and converted into a file format that is understandable by printing machines. The part is created by melting the powder using a laser or by injecting the polymeric materials, such as ABS and PLA, through the heated nozzle. Honeybees' combs, spider webs, and corals create very complex, lightweight, and stiff structures that served as inspiration to 3D printing. Multi-material and functionally graded materials can be printed, provided both chemistries match well in terms of adhesion, melting point, thermal conductivity, diffusivity, microstructure, amorphous phase, and intermetallic produced after heating and cooling [7]. Composite structures of bones, plant leaves, and bird wings inspired many bionic designs in the field of automobiles and the aerospace industries. Batch production, single-component production is possible with additive manufacturing, as there are no tool and die costs involved. Additive manufacturing is extremely useful in the production of prototypes and also in the medical field. Using MR scans, additive manufacturing models can be built, and the exact location of the tumor or cancer or other deadly growth can be identified before the operations, and doctors can do the surgery with ease, with a high success rate. Many orthopedic components are now 3D printed and custom-made. Table 2.4 shows the comparison between conventional and additive manufacturing processes and the benefits for the manufacture of bio-inspired design-based products or devices [8].

2.3.2 Functionally Graded Materials Manufacturing

It is inspired by naturally available materials, like bones, bamboo, tree bark, skin, dog fur, and seashells. Functional graded materials have different materials deposited with different compositions to meet various functionalities. It is a combination of hard, tough, flexible, and soft materials to meet various functions. For example, in soft robots, the end effector must be flexible and soft, so that the fragile material is handled effectively, and it must have a strength and stiffness so that the grabbed parts do not fall. This requirement is part-level, but this is also required at the micro- and macro-levels, as well, and can be done using 3D printing processes [9]. Multi-material printing is possible using additive manufacturing without any change of dies or tools. Because 3D printers have multiple nozzles or mix chambers, they are used for multi-component printing. The laser method is also used if high precision is needed. One way to make FGMs is by using multi-material 3D printers. These printers have multiple nozzles or mixing chambers that blend different materials before deposition. Another method is laser-based metal deposition, where the feedstock composition is controlled while a laser melts the surface. Non-uniform structures, like a bird's beak, can be bio-printed easily, but it is not possible using conventional machining processes, and even advanced machining processes like electro-discharge machining or electrochemical machining. Implants, mandalas, and bones are reproduced using additive manufacturing [10].

2.3.3 Lithography as a Manufacturing Process

It is a process of creating patterns on given surfaces using the combination of a mask and light. Complex biological structures can be printed using lithography. Micro, macro-, and nano-structures are possible with the help of lithography. Many complicated structures, like light diffusing structures of butterfly wings, superhydrophobic graded surfaces, textures for optical communication devices, and electronics device formation, can be done using lithography [11]. It is a technology that is borrowed from wafer fabrication techniques commonly used to manufacture semiconductor devices. Soft lithography uses soft materials and is suitable for flexible wearable electronics used in the medical field. Materials like polydimethylsiloxane are soft polymers. It is spun over the surface, and then the pattern is printed on this mask, which can be positive or negative photoresist. Then, it is exposed to UV light and etched chemically or otherwise to give the complex shapes and patterns. Many materials can be printed by simply changing the mask [12].

Soft lithography is cheap as it does not require the class 10 clean rooms needed for other semiconductor processes, such as chemical and vapor deposition or ion implantation. This makes the process cheap. Many footpads of micro- and macro-organisms can be easily printed on synthetic material using soft lithography. Water-repellent surfaces using the lotus effect, self-cleaning, self-aligning surfaces/structures in tissue engineering can be fabricated using soft lithography. Microchannels similar to those observed in blood vessels or plant veins can be reproduced using soft lithography. Many plants inspired water harvesting, and the growth of cells can be effectively duplicated using soft lithography. Table 2.5 shows the use of soft lithography in bio-inspired fabrication [13].

2.3.4 Self-Assembly and Directed Self-Assembly

It is one of the powerful techniques nature adopts wherein structure is built from basic elements, such as particles and molecules. Layer by layer, something similar to a wall is built using a brick as the unit element. Self-alignment or assembly comes from physical and chemical interactive forces, bonds, or surface energy, gradients of energy fields, external fields, or stimuli [14]. Extremely complex and

Table 2.5 Applications of soft lithography in bio-inspired manufacturing.

Sl. no.	Principle adopted	Examples of where they were inspired from	Practical application
1	Adhesive structures, graded structures	Gecko feet	Vertical wall-climbing robots
2	Water repulsion	Lotus effect on lotus leaves	Paints, self-cleaning textile
3	Fluid transportation	Blood vessels and plant veins	Bio sensors and a microfluidic on-board chip.
4	Cell growth	Bones	Scaffolds
5	Optical surfaces	Butterfly wings	Anti-reflective and color-changing coating used in military applications

precise structures are possible with minimum energy expenditure with this method. This method is inspired by how proteins from cell membranes perform protein folding. This method is a bottom-up fabrication process in the sense that it starts from molecular, micro-, macro-, or nano-level structures, and it is not a subtractive or cutting process.

This process uses block copolymers, which are a bunch of different polymers fastened together. They can self-align into dots, lines, and holes. The first thin film of a copolymer is deposited on the surface using any spinning process. Using lithography or soft lithography, patterns are generated on the substrate. When a copolymer is exposed to heat or a special solvent, it will mobilize molecules, and the molecules self-organize and take the shape of the pattern required on the surface [15]. The advantage is that the existing semiconductor fabrication method can be used and there is no necessary for masks in these processes. Directed self-assembly (DSA) can also be used to fabricate nano-patterned surfaces on membranes. These

Table 2.6 Applications of DSA in bio-inspired manufacturing.

Sl. no.	Principle adopted	Example from where they were inspired	Practical application
1	Antibacterial surfaces	Shark skin	Medical polymers and devices
2	Colors without pigments	Butterfly wings	Color sensor and optical devices
3	Molecular patterns	Cell membranes	Sensors in a chemical factory
4	Gradient surfaces	Beetles	Water harvesting systems
5	Color changing	Butterfly wings	Anti-reflective and color-changing coating used in military applications
6	Lightweight structures	Spider silk	Synthetic fibers that are stronger than steel

members are used as sensors in chemical industries to detect the minute changes in the fluid or gas environment. Both organic and non-organic materials can be used in this method, which makes it more versatile. When producing colors without pigments, such as butterflies, DSA is useful in displacing [16]. Table 2.6 provides an application of DSA used in bio-inspired manufacturing.

2.3.5 Bioprinting

It is an advanced fabrication process that makes use of cells and biomaterials, which are made layer by layer. A precise arrangement of cells, biomaterials, and growth substances is introduced to create a structure, such as skin tissue, vascular tissue, and cartilage. This method is inspired by the skin repair mechanism in humans and regenerative processes in many organisms. This method is very useful in tissue engineering, regenerative organs, implants, personalized drug delivery, and medicines [17]. Many inkjet processes, extrusion processes, laser laser-assisted processes have been developed based on the inspiration obtained from embryonic development and tissue repair mechanisms in animals and organisms. There are three types of bioprinting. The cell is mixed with polymer, and bio-link/bio-ink is formed. Then, using Inkjet bioprinting, extrusion, or laser-assisted processes, it is printed and finally implanted into the required animal or human. The process is shown in Figure 2.3.

Inkjet bioprinting uses bioink produced by a combination of cell, polymer, cross-linker, and growth factor, and deposits drop by drop onto a surface using an actuator, either thermal or piezoelectric, and is used for printing low-viscosity materials. Extrusion bioprinting also uses bioink, and it is extruded using the nozzle [18]. Complex structures, such as skin and cartilage, can be produced using this process. Laser-assisted systems use a laser beam to send the droplet onto the substrate and provide high resolution and cell adherence. Octopus arms and insect legs inspired bio robotics development. Researchers can print a vascular network and an efficient nutrient transport system using bio printing [19]. Bio printing involves four

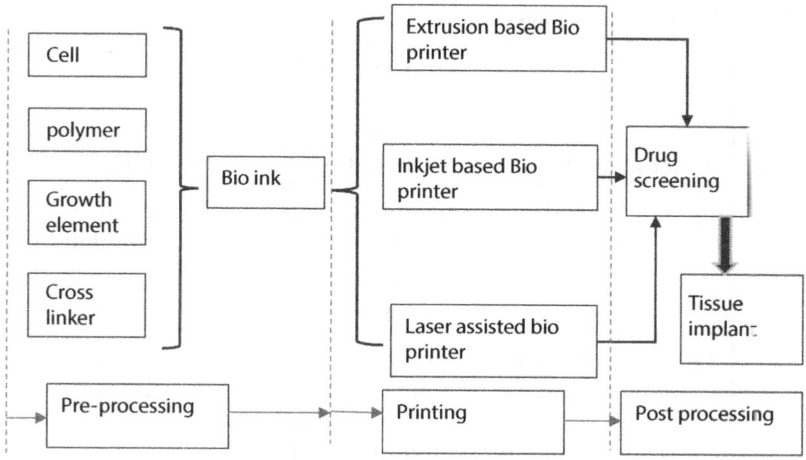

Figure 2.3 Bio printing processes and stages in bio printing.

important processes: pre-processing, biolink preparation, printing, and post-processing before implant, as shown in Figure 2.3.

2.3.6 Microfluidics as a Fabrication of Bio-Inspired Systems

Microfluidic fabrication is one of the advanced fabrication technologies that uses properties of fluid and its behavior on a microscale. Many microfluidic devices are designed and manufactured by taking inspiration from blood flow in veins in humans, nutrient transportation mechanisms in plants, and enzyme secretion [20]. Microfluidic devices work with high precision in different environments. Artificial leaves, biosensors, synthetic blood vessels, and heart valves are a few examples. They are also part of lab-on-a-chip devices, where many functions can be combined in one for improving overall efficiency.

Microfluidic devices are generally fabricated using wafer fabrication methods known as photolithography processes. The CAD software is used to make a detailed layout, and a master mold is created using photolithography. Then, polymers, like Polymethysiloxane, are cast over the mold prepared earlier [21]. Micro channels are prepared through curing processes and then finally bonding the system to the chip or the substrate. Femtosecond laser processes are

Table 2.7 Use of various materials.

Sl. no.	Material used	Properties that make them suitable for fabrication	Suitability
1	Epoxy-based (SU-8)	Negative photoresist with good sensitivity and good resolution	Molds and bio MEMS and semiconductor devices
2	Polydimethylsiloxane	Synthetic polymer, flexible, thermally stable, biocompatible, and transparent	Flexible and wearable devices
3	Glass	Inert, transparent	Image and chemical analysis
4	PMMA Polymethyl Methacrylate	Rigid but easily machinable	Used for micro channels
5	Optical surfaces	Butterfly wings	Anti-reflective and color-changing coating used in military applications

nowadays used to make precision microfluidic channels. Various materials used and their suitability can be found in Table 2.7.

Advantages of Microfluidic Fabrication:

- Fluid dynamics in a microchannel can be controlled well.
- Wastage is minimal with less energy consumption
- Efficiency is high
- Easy to integrate with actuators and sensors
- Easy to prototype and manufacture due to its small size.

Disadvantages:

- Expensive as it requires a clean room facility, which is highly cost-intensive
- There will be chemical reactions between the polymer and chemicals
- Because of the size, it is difficult to remove clogging, if any
- Difficult for scaling up as fluid dynamics at the macro scale will be different, and linear scaling and fabrication may not work and give the desired results.

2.3.7 Bio Mineralization

It is the way by which animals and organisms produce structures made of minerals like corals, seashells, snails, tortoise shells, and bones under ambient conditions. Ceramic specialists are inspired by these ceramic production processes at low temperatures [22]. In conventional methods, ceramics are produced by high-temperature sintering processes, which are highly energy-consuming processes. Researchers used by-products obtained from a certain class of bacteria known as *Sporosarcina pasteurii*. These bacteria can produce special calcium carbonate, which can enrich soil and be used as a good substitute for the repair of building cracks [23].

Advantages of Biomineralization-Based Fabrication:

- Most of the formation is at room temperature and pressure.
- Great mechanical properties.
- Shape, size, and curvature can be controlled easily.
- Biocompatible processes.
- Less energy-consuming processes, so the efficiency of conversion is high.
- Process integration is not difficult, unlike photolithography.

Disadvantages:

- The process of formation takes time since it occurs naturally.
- Requires a controlled atmosphere, both chemical and physical.
- Large-scale production is difficult.
- Minerals available for fabrication are limited.

2.3.8 Origami-Inspired Design

It is based on the simple art of paper folding, which is inspired by leaves, insects, wings, proteins, and many tissues available in organisms. With the advent of microscopy, researchers can invent more and more in this field, especially for medicine, military applications, and space applications. Unfolding solar panels for aircraft, foldable houses used during calamities, medical stents, flexible wearable electronics, and origami boats are a few examples [24]. Design is quite compact since it is foldable without many complicated actuators, sensors, motors, and controllers. It is lightweight, adaptable, and easily transportable from place to place. One can easily program the ability of material to fold by the following patterns. These patterns act as interlocks [25]. Polyimide films, silicone elastomers, hydrogels, and certain shape memory alloys are bio-inspired materials. Laser etching, patterning, inkjet printing, and additive manufacturing, discussed above, are used to produce these types of products. Wearable electronics, flexible electronics, tissue engineering, space, and soft robotics are the areas in which these principles are put into practice [26].

Advantages of Origami-Inspired Fabrication:

- Very compact and deployable anywhere easily.
- Mechanical complexities and parts are minimal.
- Materials are soft and foldable.
- Shapes and functions can be controlled dynamically.

- A graded scale from micro to nano is possible.
- Storage is easy, and the transportation and installation costs are lower.

Disadvantages:

- If folded many times, the part will fail by fatigue.
- The pattern is quite complex and not easily relatable.
- Materials available for this application are limited.
- Cost-intensive especially for miniature devices.

2.3.9 Bioelectronic Interfaces

It is hybrid manufacturing processes that combine the advantages of lithography, bioprinting, and wafer fabrication deposition techniques. Here, biological tissues are integrated with electronics to mimic the superior characteristics of flexibility, responsiveness of natural tissues available in nature [27]. This type of design is inspired by the flow of electrical neuron signals in our nervous systems, which involves both electronics and tissues. Electronic skin, sensors, and neural implants are a few examples that use hybrid fabrication processes. It is the integration of electronics, soft and hard materials, and the responsiveness of natural tissues. Fabrication of such systems requires methods that allow integration of hard and soft materials, often at the micron or submicron level, which is made possible through hybrid approaches combining lithography, printing, and deposition techniques. Here, interfacing between the tissues, signal processing through electronics, and materials are needed for efficient functioning [28]. Lithography, inkjet printing, and additive manufacturing are needed in addition to deposition processes like chemical vapor deposition, which can deposit thin film/layer conductive or nonconductive material, like in semiconductor wafer fabrication. Conductive polymers, carbon nanomaterials, elastomers, hydrogels, metal nanostructures, wires, and particles are used in this process.

Advantages of Bioelectronic Interfaces:

- Can combine both conductivity and flexibility on one device
- Skin, nerve, and organ integration is possible, and they work in unison
- Real monitoring of parameters is possible
- Signal is detected with high resolution and good repeatability
- Wearable and life-supporting devices are possible to integrate.

Disadvantages:

- Both phase integration of hard and soft materials is difficult
- Compatibility and durability can be an issue
- Interference and noise during signal transport or processing
- Fabrication cost is high
- Limited to intermittent operations due to the unavailability of the source supply.

2.4 Materials and Material Systems in Bio-Inspired Manufacturing

Although many materials are available in nature, secondary operations, selection, synthesis, and processing are needed to get the final device. The choice of materials for a particular bio-inspired design is important to give the required function and performance index [29]. Different applications demand different properties and structures. Therefore, a variety of materials are needed. Some materials are combined as in functionally graded materials, and some materials are to be reinforced as in the case of composites, some materials are to be interfaced as in bio sensors, and some materials are to be of different structures. Material selections depend on required properties like weight, resilience, stiffness, toughness, malleability, compatibility, adaptability, flexibility, conductivity, and dynamic response [30]. Spider silk is as strong as steel and resilient. Nacre is tough, and plants show an extraordinary dynamic response. Nature offers a wide variety of devices, products, and materials that have specific functions or groups of functions with exceptional performance and efficiency in terms of using natural resources and energy [31, 32].

The superstructure of the nacre shell consists of aragonite platelets, which are brittle in nature but arranged in a mortar–brick with an organic reinforcement. Although aragonite is brittle in nature, its arrangement in nacre makes it a tough material. Mortar and brick structures are such that they can close the crack or cause deflection of the crack and good surface characteristics to dissipate the energy in the most efficient way [33]. This is mimicked in aerospace structures and armor panels, which have fracture resistance and are highly durable. Spider silk is produced by a protein-based compound. It has high tensile strength with low defects, making it highly tough and elastic. It is stronger than steel for the same diameter and length. It is biocompatible, and the spider makes it room temperature and pressure, so it is a highly energy-efficient manufacturing process. Researchers have mimicked spider silk and produced remarkable textile materials, flexible and wearable electronics, and sutures [34–36]. The DNA technology and microbial fermentation are two techniques researchers have used to develop synthetic spider silk.

The structure of a plant cell is porous and has reinforcement such that it is fracture and fatigue-resistant. It is made of lignin, cellulose, and hemicellulose and provides stiffness and toughness to the structure. Researchers have developed polymer and metallic foam based on this concept. Natural structures are always biodegradable, environmentally friendly, and sustainable. Nano-cellulose structures are built, which are very lightweight, superior in stiffness, strong, yet flexible like plant cell-based materials [37]. Bone is another natural material from humans, animals, birds, and even some organisms. Bone is a composite material with exceptional resilience and strength, which is a combination of a mineral phase called hydroxyapatite and a matrix phase called flexible collagen. It is a strut-like structure. The structure in bone is a graded structure from macro- to nanoscale [38]. Many implants and prosthetics were developed through additive manufacturing, mimicking the structure of bone. Chitosan is another biodegradable material that is widely used in medical applications. Biopolymers are commonly used for dressing wounds and as a filtration membrane. There are many types of thin-film-based membrane structures used in chemical engineering to mimic the exoskeleton of a beetle for the design and development of devices and sensors.

2.5 Conclusion

Bio-inspired manufacturing was once only a research domain. Now, it has entered the industrial domain. Aerospace industries developed many bio-inspired products, devices, and structures. In some cases, weight reduction of up to 30 % was possible without compromising other properties [39]. The medical field too incorporated many bio-inspired designs to make the products, devices, and prosthetics, which have increased the life span of human beings. There are many bio-inspired environmental applications, like filtration systems, desalination, underwater repair, and marine coatings. Synthetic biology and genetic engineering are gaining lots of potential in the field of bio-inspired design, even pushing the natural boundaries of bacteria, viruses, etc. Plastics are causing greater harm to society and the environment, so researchers are developing biodegradable plastics using many avenues, such as yeast-based alternatives. Protein-based foams are being researched in many engineering fields to develop a better structure. Another new technology is 4D printing. Bio printing is also available for better product design and manufacture [40]. Research is going on advanced response stimuli materials for medical applications and in aerodynamics. Swarm manufacturing is adopted slowly in robot industries for self-organizing, local decision-making, self-programming using artificial intelligence, and extra-terrestrial applications. A lot of research is being done to develop cognitive materials with advanced signal processing, self-networking, self–healing, and stimuli-responsive capabilities. Responsive clothing is another possibility. Many progresses are on the way, and it is important to address ethical issues, ecological balance, environmental impact, and natural balance before unlocking revolutionary bio-inspired design for humankind.

References

1. Vincent, J.F.V., Bogatyreva, O.A., Bogatyrev, N.R., Bowyer, A., Pahl, A.-K., Biomimetics: Its Practice and Theory. *J. R. Soc. Interface*, 3, 9, 471–482, 2006.
2. Bhushan, B., Biomimetics: Lessons from Nature–An Overview. *Philos. Trans. R. Soc. A Math. Phys. Eng. Sci.*, 367, 1893, 1445–1486, 2009.
3. Bar-Cohen, Y., Biomimetics—Using Nature to Inspire Human Innovation. *Bioinspir. Biomim.*, 1, 1, P1–P12, 2006.
4. Fratzl, P. and Barth, F.G., Biomaterial Systems for Mechanosensing and Actuation. *Nature*, 462, 7272, 442–448, 2009.
5. Dicker, M.P.M., Rossiter, J.M., Bond, I.P., Weaver, P.M., Biomimetic Photo-Actuation. *Adv. Mater.*, 26, 5, 703–709, 2014.
6. Liu, Y., Boyles, J.K., Genzer, J., Dickey, M.D., Self-Folding of Polymer Sheets Using Local Light Absorption. *Soft Matter*, 8, 6, 1764–1769, 2012.
7. Zolfagharian, A., Kaynak, A., Khoo, S.Y., Kouzani, A.Z., Nahavandi, S., 3D Printed Soft Actuators: Design, Fabrication, and Control. *J. Micromech. Microeng.*, 26, 10, 103001, 2016.
8. Lee, A., Hudson, A.R., Shiwarski, D.J., Tashman, J.W., Hinton, T.J., Yerneni, S., Bliley, J.M., Campbell, P.G., Feinberg, A.W., 3D Bioprinting of Collagen to Rebuild Components of the Human Heart. *Science*, 365, 6452, 482–487, 2019.
9. Muth, J.T., Vogt, D.M., Truby, R.L., Mengüç, Y., Kolesky, D.B., Wood, R.J., Lewis, J.A., Embedded 3D Printing of Strain Sensors within Highly Stretchable Elastomers. *Adv. Mater.*, 26, 36, 6307–6312, 2014.
10. Rajabi, H., Darvizeh, A., Shafiei, A., Taylor, D., Dirks, J.-H., A Comparative Study of the Effects of Bio-Inspired Scales on the

Mechanics of Flexible Structures. *J. Mech. Behav. Biomed. Mater.*, 60, 232–242, 2016.

11. Kim, Y., Yuk, H., Zhao, R., Chester, S.A., Zhao, X., Printing Ferromagnetic Domains for Untethered Fast-Transforming Soft Materials. *Nature*, 558, 7709, 274–279, 2018.
12. Martinez, R.V., Branch, J.L., Fish, C.R., Jin, L., Shepherd, R.F., Nunes, R.M., Suo, Z., Whitesides, G.M., Robotic Tentacles with Three-Dimensional Mobility Based on Flexible Elastomers. *Adv. Mater.*, 25, 2, 205–212, 2013.
13. Kalantari, A. and Spinks, G.M., Bioinspired Soft Robotics: Materials, Actuation, and Applications. *Adv. Intell. Syst.*, 3, 2, 2000188, 2021.
14. Jiang, Y. and Liu, Z., Soft Bioinspired Robotics Driven by Liquid Crystal Elastomers. *Adv. Funct. Mater.*, 32, 13, 2107414, 2022.
15. Sydney Gladman, A., Matsumoto, E.A., Nuzzo, R.G., Mahadevan, L., Lewis, J.A., Biomimetic 4D Printing. *Nat. Mater.*, 15, 4, 413–418, 2016.
16. Sharma, A., Kaur, A., Pal, A., Bioinspired Micro/Nanostructures for Multifunctional Applications. *J. Micromech. Microeng.*, 33, 2, 023002, 2023.
17. Ma, W., Pang, H., He, D., Liu, H., Hu, Z., Nature-Inspired Materials and Structures for Flexible Electronics. *npj Flex. Electron.*, 7, 1, 7, 2023.
18. Truby, R.L. and Lewis, J.A., Printing Soft Matter in Three Dimensions. *Nature*, 540, 7633, 371–378, 2016.
19. Chen, T., Bilal, O.R., Shea, K., Daraio, C., Harnessing Buckling to Design Architected Materials That Exhibit Effective Negative Swelling. *Adv. Mater.*, 30, 49, 1802405, 2018.
20. Wehner, M., Truby, R.L., Fitzgerald, D.J., Mosadegh, B., Whitesides, G.M., Lewis, J.A., Wood, R.J., An Integrated Design and Fabrication Strategy for Entirely Soft, Autonomous Robots. *Nature*, 536, 7617, 451–455, 2016.
21. Song, S., Ke, Y., Ma, Z., Liu, J., Liu, Y., Liu, L., Bioinspired Structures and Smart Materials for Soft Robotics: Progress, Challenges, and Future Directions. *Adv. Mater. Technol.*, 8, 2, 2200900, 2023.
22. Zhang, Y., Wang, J., Wang, Y., Recent Advances in Nature-Inspired Materials for Energy Storage and Conversion. *Nano Res.*, 16, 2, 2747–2770, 2023.

23. Liu, D., Li, Z., Yang, Y., Chen, C., Bio-Inspired Structural Design in 3D Printing: A Review. *Addit. Manuf.*, 59, 103167, 2023.
24. Wu, C., Ma, B., Li, Z., Li, Q., Hierarchical Bioinspired Structures for Multifunctional Applications. *ACS Nano*, 16, 1, 345–361, 2022.
25. Chen, Y., He, X., Du, C., Zhang, Q., From Nature to Engineering: Multiscale Bioinspired Hierarchical Materials. *Adv. Mater.*, 34, 6, 2103953, 2022.
26. Zhang, H. and Wang, Z.L., Bioinspired Triboelectric Nanogenerators for Energy Harvesting. *Adv. Funct. Mater.*, 31, 23, 2100400, 2021.
27. Lin, Y., Gao, M., Wang, C., Design of Bioinspired Smart Surfaces with Switchable Wettability for Functional Applications ACS. *Appl. Mater. Interfaces*, 13, 6, 7486–7504, 2021.
28. Li, Y., Li, J., Yang, H., Biomimetic Design and Fabrication of Functional Surfaces for Water Harvesting and Self-Cleaning. *Adv. Mater. Interfaces*, 7, 14, 2000450, 2020.
29. Wang, Q. and Wang, Z., Bioinspired 3D Printing of Self-Healing Soft Materials for Wearable Devices. *Adv. Mater. Technol.*, 5, 11, 2000536, 2020.
30. Wang, Y., Liu, L., Zhan, J., Bioinspired Nanostructures for Superhydrophobic and Antifouling Surfaces. *J. Mater. Chem. A*, 7, 3, 1167–1189, 2019.
31. Jiang, L. and Zhao, Y., Bioinspired Multiscale Surfaces for Energy and Environmental Applications. *Nano Energy*, 47, 203–231, 2018.
32. Zhao, H., Xi, J., Tang, Z., Bioinspired Design Principles for Multifunctional Surfaces and Interfacial Materials. *Adv. Mater. Interfaces*, 4, 2, 1600491, 2017.
33. Zhu, Y., Liu, Y., Hu, J., Bioinspired Design of Soft Actuators Based on Liquid Crystal Elastomers. *Macromol. Rapid Commun.*, 38, 24, 1700329, 2017.
34. Zhang, H., Xia, J., Wang, P., Biomimetic Anti-Fog Coatings: Recent Advances and Emerging Applications. *Chem. Soc. Rev.*, 45, 3, 315–338, 2016.
35. Lu, Y., Zhang, S., Liu, X., Bioinspired Self-Healing Superhydrophobic Coatings. *Adv. Mater.*, 27, 44, 7614–7619, 2015.
36. Zhang, Y. and Zhao, Y., Bioinspired Light-Responsive Materials for Smart Systems. *Chem. Rev.*, 114, 9, 4961–4980, 2014.

37. Li, C., Lv, J., Zhu, J., Self-Healing Bioinspired Gels for Tissue Engineering and Soft Robotics. *ACS Appl. Mater. Interfaces*, 5, 19, 9040–9047, 2013.
38. Liu, Z., Wang, Y., Xie, T., Bioinspired Design and Applications of Hydrogels. *Macromol. Chem. Phys.*, 213, 12, 1170–1177, 2012.
39. Aizenberg, J. and Fratzl, P., Biomimetic Materials and Design: Biointerfacial Strategies, Tissue Engineering and Targeted Drug Delivery. *Adv. Mater.*, 21, 38–39, 3870–3874, 2009.
40. Sanchez, C., Arribart, H., Giraud Guille, M.M., Biomimetism and Bioinspiration as Tools for the Design of Innovative Materials and Systems. *Nat. Mater.*, 4, 4, 277–288, 2005.

3

Bio-Inspired Computation: Leveraging Ant Behavior for Smart Algorithm Design

Abstract

Most algorithms artificially made by humans suffer from some serious drawbacks, such as the large-scale computational power requirement when scaled to large systems or increased complexity, among others. To solve these problems, several resources have suggested the idea of looking into natural systems for inspiration. Since most natural systems are the result of thousands of years of evolution, they tend to be highly optimized, offering insights for designing efficient algorithms. One such category of organisms that provides numerous ideas is ants. One significant advantage of studying ants is that most of their behaviors can be easily modeled, as they do not have complex brain structures like humans. Instead, their actions are largely driven by chemical or mechanical stimuli rather than complicated thought processes. Additionally, the intricate structures of ant societies should also be noted for their relevance to engineering and optimization. In this paper, several algorithms based on different types of ant behaviors are studied. It is important to note that most of these algorithms are not exact replicas of how ants behave, but rather mimic their behavior to extract the essential characteristics for achieving specific goals. In the initial sections, several path-planning algorithms are discussed, commonly referred to as ant colony optimization (ACO), which remains a prominent area of research. Following this, various algorithms inspired by the complex social structures of ants are presented under the title of swarm intelligence algorithms. These include dynamic task allocation algorithms and decentralized control strategies. Additionally, various other algorithms, namely, hidden Markov and link prediction, are also described that are used for the next prediction using the ACO. Toward the end, some of the fascinating applications of these algorithms are highlighted. These applications include swarm robotics, such as a swarm of drones, machine scheduling systems for optimized task allocation, and tool path optimization in additive manufacturing to reduce transition time and improve surface quality.

Jagadeesha T. and Sandip Kunar. Bio-Inspired Manufacturing Systems, (57–84) © 2026 Scrivener Publishing LLC

***Keywords*:** Ant colony optimization, bio-inspired computing, swarm intelligence, task scheduling, additive manufacturing

3.1 Introduction

Nature is an excellent source of inspiration, and one such source is the behavior of ants. Ants use unique strategies to establish colonies, communicate with one another, and even interact with other organisms, allowing them to survive and coexist. These behaviors provide methods for solving complex problems. For example, ants have inspired solutions to problems like the Traveling Salesman Problem (TSP), sorting challenges, and prediction tasks—problems for which traditional human algorithms often require significant computational time and produce less efficient results. Ant colony optimization (ACO) is a computational technique that mimics the foraging behavior of ants to find optimal solutions [1]. Ants exhibit several behavioral traits that allow them to leave pheromone trails, communicate indirectly, and follow the shortest paths based on pheromone concentration and visibility, adjusted by evaporation rates, as shown in Figure 3.1. These behaviors can be broken down into simple steps, which can solve a wide range of complex problems.

Algorithms inspired by ACO are widely used in various applications, including path optimization, job allocation, and swarm robotics. In swarm robotics, for instance, ACO facilitates communication and coordination between robots. These natural approaches provide solutions for optimization problems in various fields [2]. Subsequent sections deal with various algorithms inspired by ACO and examine their applications in real-world scenarios.

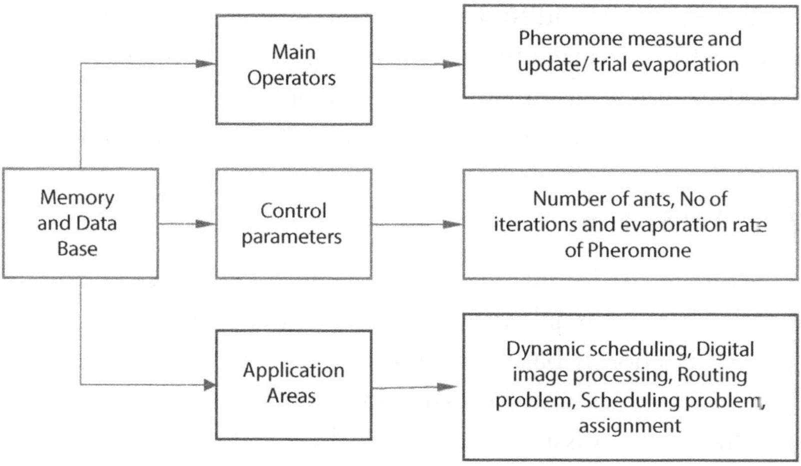

Figure 3.1 Ant algorithm optimization.

3.2 Algorithms

3.2.1 Path Planning Algorithms

The TSP involves finding the shortest path within a time-bound period for connecting all the cities for travel. Traveling salesman problems are normally defined based on the number of cities visited. The foraging behavior of ants inspires countless meta-heuristic path planning algorithms. However, there are certain issues with the algorithms explained, other than those inspired by natural habitats. There are no known polynomial solutions for the existing problems, along with the exponentially increasing number of paths. Real-time problems become even more complicated in addressing situations like current traffic and volume spikes.

While moving to and from the food source, ants deposit a chemical substance known as a pheromone along the path. These pheromones act as stimuli for other ants to collect the food. That is the ants' sense pheromone, and they prefer to go along a path of the highest pheromone concentration. Also, the pheromone undergoes continuous evaporation into the environment. As time passes, the more successful paths (where ants easily get food and return) become dense with pheromone, and those paths emerge as the optimum path from the ant nest to the food source. Inspired by this, a set of algorithms is proposed to find near-optimal paths to a destination. They can also be used to find solutions for any other optimization problems [3]. The major algorithms are discussed in the following sections.

3.2.2 Ant System

This is one of the initial algorithms proposed on ACO. In this algorithm, a set of artificial ants is made to find a path for the problem. The path is chosen based on a stochastic mechanism using both pheromone values and heuristic information to choose the next node when the ant is at a given node. The equation for the stochastic mechanism is given as:

$$p_{ij}^{k} = \begin{cases} \dfrac{\tau_{ij}^{\alpha}\eta_{ij}^{\beta}}{\sum_{c_{ij}\in N(s^{p})}\tau_{ij}^{\alpha}\eta_{ij}^{\beta}} & \text{if } c_{ij}\in N(s^{p}), \\ 0 & \text{otherwise.} \end{cases} \tag{3.1}$$

Where p_{ij}^{k} is the probability of selecting a node. If the node is previously visited, the value is set to zero (One city should be visited only once for the TSP).

After a set of m ants find their path (one iteration), the pheromone concentration along all the edges is updated as:

$$\tau_{ij} \leftarrow (1-\rho)\cdot\tau_{ij} + \sum_{k=1}^{m}\Delta\tau_{ij}^{k} \tag{3.2}$$

Where ρ is the constant related to evaporation, $\Delta\tau_{ij}^{k}$ is the pheromone concentration on edge ij provided by the kth ant. The value of the pheromone is given by:

$$\Delta\tau_{ij}^{k} = \begin{cases} \dfrac{Q}{L_k} & \text{if ant k used edge (i, j) in its tour,} \\ 0 & \text{otherwise.} \end{cases} \tag{3.3}$$

Where Q is a constant and L_k is the total length of the path found by that particular ant [4].

3.2.3 MAX–MIN Ant System

This is a modification of the initial 'Ant System.' The key differences are:

- Minimum and maximum values are given for the pheromone level a path can have. Otherwise, a path might not be considered just because the initial ants did not choose it.
- Only the best ant (Can be best in all iterations combined) can modify the pheromone values. So the ants who chose the unsuitable path cannot update the pheromone, and the system will reach the optimum faster. But the exploratory nature of the system to new paths might be compromised.

In the Ant Colony System Algorithm, in addition to these, the pheromone values are also updated within an iteration in such a manner that following ants will choose a path that is different from

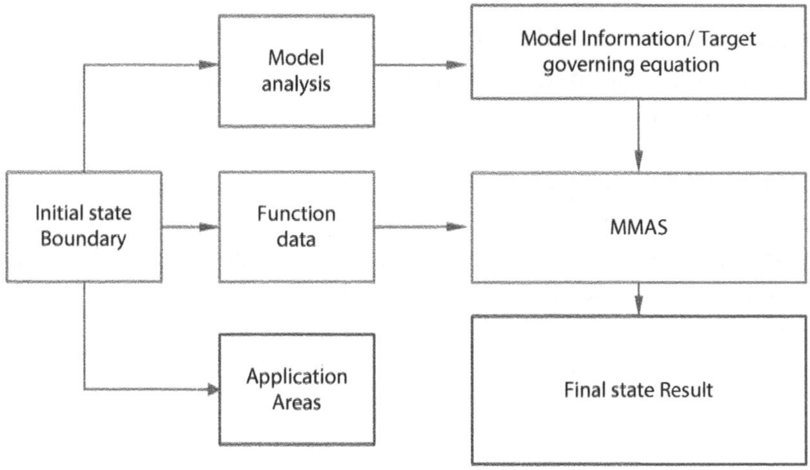

Figure 3.2 Max–Min ant system.

the previous ants in the current iteration [5]. The configuration is given in Figure 3.2.

3.2.4 AddACO

The ants are placed randomly in a graph and are allowed to move randomly to any location to complete their tour. The randomness in the future state depends only on the current state and the transition could be marked as:

$$p_{n_j \to n_i}(t) = \begin{cases} \dfrac{\sum_{b \in B} \alpha^b \, \varphi_{e_{ij}}^b(t)}{\sum_{e_{ik} \in E,\, n_k \in V^m(a)} \varphi_{e_{ik}}^b(t)}, & \text{if } n_j \in V^m(t); \\ 0, & \text{otherwise,} \end{cases} \tag{3.4}$$

where p_n is the transition from one state to another with $V^m(t)$ is a set of all the points not visited by ants for the mth ant. B is the set of behavioral traits for ants.

The model proposed suggests a canonical multiplicative form. The ants use several iterations to complete the tour to the intended cities. When one ant sets out on the tour to the town, it returns to the initial node. Each time the destinations are reinitialized and the total distance traveled is evaluated, the minimum path is selected. The algorithm is allowed to repeat an arbitrary number of times. This procedure continues till the ants trace the same path or the sequence again, which is defined as the equivalent tour. With this, there are three variations for the same proposal:

AddACO-V1: This is similar to all the ACO classic problems, which assume the decision taken by the ant to choose a path to be either to travel by the heuristic path or to follow the path that has the highest concentration of pheromones, for which the equation could be modified as:

$$p_{n_i \to n_j}(t) = \begin{cases} \alpha \dfrac{q_{e_{ij}}(t)}{\sum_{e_{ik} \in E, n_k \in V^m(t)} q_{e_{ik}}(t)} + \beta \dfrac{(l_{e_{ij}})^{-1}}{\sum_{e_{ik} \in E, n_k \in V^m(t)} (l_{e_{ik}})^{-1}}, & \text{if } n_j \in V^m(t); \\ 0, & \text{otherwise,} \end{cases} \tag{3.5}$$

where $q_e(t)$ indicates the amount of pheromones left behind in a given time t on the edge e. The length of the path is denoted by l_e.
AddACO-V2: In this model, the ant is given a randomness in its behavior; noise is added into its motion. In such a case, the alignment of the track taken by the ant also plays a role while making the decision.
AddACO-V3: The third model brings in the concept of inertia. Ants prefer to move along the same path without any change in direction.

The influence of pheromone trail is included in all the above-mentioned algorithms and assumes spatiotemporal rule is same for all the above three. In further detailed versions uses variations of patterns to occur at every N^{th} iterations of the tour. The first theorem explained could be applied to any scenario but the computation time is comparatively larger for which the time could also be not predicted [20].

3.3 Swarm Intelligence Algorithms

Swarm intelligence is an algorithm inspired by the idea of ant colonies, where they have a task in which different parts are solved by different groups, making the problem-solving much more efficient by utilizing the group's combined knowledge. Everyone in the group should be aware of the others' jobs and should have a means of communication among them. Various ideas were inspired by nature for solving such problems, among which is the relationship between aphids and ants, who have lived together for a very long time [19].

3.3.1 Aphid-Ant Mutualism

The worker ants in the ant family work hard and travel miles to prepare food. The ant colony also depends on the Aphids, who help them in searching for food, and the ants help protect the aphids in return. This relation is called mutualism.

Aphid-ant mutualism (AMM) creates a random population, which is then randomly split into aphids and ants, forming a colony with a greater number of aphids than ants. Ants and aphids interact and benefit from each other: ants help protect aphids, while aphids assist in the food search. The entire process takes place in four steps. The first step involves generating a random population, with aphids seeking vegetables that have more sap [25]. The strongest aphid has a greater chance of reaching the biggest plant, while weaker aphids move towards the colonies. In the second phase of mutualism, aphids and ants work together to discover better food sources. In the final phase, aphids grow heavier and fly towards the next ant colony, helping them find better food sources.

The algorithm begins by creating a population and dividing it randomly into ants and aphids, with the number of aphids greater than the number of ants. Eventually, both will form separate colonies for mutual benefit. Once the colonies are formed, ants and aphids use different methods for searching and sorting. If they both move in the same direction, they will get stuck somewhere [28]. The position of the ants depends on the selected aphids, based on the pheromones. The optimization process controls the movement of the ants towards the selected aphids. Depending on the evaporation rate, the amount of pheromone varies. In the initial steps, the ants explore the area between the selected aphids and their current position. In the later stages, they focus on taking advantage of the good spots near the best aphids.

The division of the population into ant and aphid can be represented as:

$$Aphid_{k,j} = LB_j + rand \times \left(UB_j - LB_j\right), \quad k = 1,2,\ldots,n, \tag{3.6}$$

$$Ant_{p,j} = LB_j + rand \times \left(UB_j - LB_j\right), \quad p = 1,2,\ldots,m. \tag{3.7}$$

where LB_j and UB_j represent the lower bounds and the upper bounds, respectively. $rand$ a random number between 0 and 1.

$Aphid_{k,j}$ represent the aphid's position and $Ant_{p,j}$ shows the ant's position.

The positions of ants and aphids are updated very frequently. The aphids, which are the strongest among all, independently search for better locations and food. Some of the weak aphids remain along with the ants and assist them in the food search. The position for the weakest among them is not updated, and their position depends on the position of aphids on whom they depend.

$$Aphid_{i,j}(t+1)^{\eta} = Ant^{\eta}_{rand(t)} + \beta \times rand \times \left| Aphid_{i,j}(t)^{\eta} - Ant^{\eta}_{rand(t)} \right| \tag{3.8}$$

Here, i represents the position of the aphid in j^{th} dimension and $t+1$ represent the number of iterations in the nth colony.

Next, the ant's position is always determined concerning the strongest aphid, as they always move towards the strongest aphid in search of a better area and food. The movements of ants are based on the number of pheromones, which is dependent on another factor called the evaporation rate [26]. Ants always move in a path with the greatest concentration of pheromones. Furthermore, aphids can fly towards the adjacent colony in a spiral motion.

The experimental results show that AAM gives a better result with a better convergence rate, along with a smaller number of iterations, giving it a smaller computational time when compared to the other algorithms at a lower cost [24].

3.3.2 Decentralized Control Algorithm for a Collection of Robots

Swarm intelligence algorithms often involve multiple agents collaborating to achieve a common task. Unlike the previously discussed algorithms, this approach emphasizes cooperation among agents. Such tasks, termed collective tasks, can be categorized into cooperative and non-cooperative tasks [6]. Cooperative tasks require the combined effort of multiple agents, as an individual cannot complete

Table 3.1 Behavioral framework and robot programming with five behaviors.

Find (default behavior)	Robots move randomly until they detect another robot.
Follow	Activated when a robot senses another robot, it moves towards the detected robot.
Slow	When close to another robot, the robot reduces its speed to avoid collisions.
Avoid	Activated to prevent collisions when two robots are too close.
Goal	When an open space near the box is detected, the robot moves to that spot and pushes the box forward.

them alone. For example, ants collectively carrying a large insect demonstrate cooperative behavior.

In robotic systems, decentralized control algorithms have been proposed to eliminate the reliance on large centralized systems, such as those used in warehouse automation. These algorithms enable a group of simple robots to function collectively as an intelligent system. One application is box-pushing, where a group of robots locates and moves a heavy box requiring more force than an individual robot can generate. The Behavioral Framework is given in Table 3.1.

By following these behaviors, the robots collectively complete the task of moving the box without requiring individual intelligence.

Challenges and Solutions

Despite its simplicity, this algorithm has potential issues. For instance, if most robots gather at the wrong side of the box, they may push it backward. To address this, a centralized system monitors the direction of the box's movement [6]. If the box moves backward continuously for a specific duration, the system resets the robots to random movement. This increases the probability of the robots reassembling in positions that push the box forward.

Table 3.2 Additional requirements.

Dynamic Task	Implementing dynamic task allocation to ensure balanced robot distribution around the box.
Detecting	Introducing mechanisms for detecting and addressing inefficiencies in real-time.
Communication protocol	Adding communication protocols for better coordination among robots.

Future Improvements

Additional features may be required to enhance the algorithm's reliability, adaptably, and contextuality as given in Table 3.2.

This decentralized control algorithm exemplifies how simple rules and local interactions can lead to emergent intelligent behavior in robotic systems [7].

3.4 Algorithm for Collective Sorting Robots

As mentioned in the previous algorithm, certain tasks can only be completed through the cooperation of multiple agents, while others can be accomplished individually but are more efficient when performed by a group. Sorting is one such task observed in ant colonies. In these colonies, brood and food are not uniformly distributed but are arranged in piles next to each other. Unlike humans, who rely heavily on mutual communication for successful sorting tasks, ants achieve this without any known mutual communication mechanisms or hierarchical organization [8]. No single ant dictates decisions; instead, simple individual behaviors lead to the desired collective outcome.

Inspired by this behavior, a simple algorithm is proposed for robotic sorting. This algorithm requires only local environmental information and short-term individual memory. It is probabilistic and was tested using a Monte Carlo simulation to explain its functionality.

3.4.1 Simulation Environment and Algorithm

The environment is modeled as a square network of points. At time zero, two types of objects are randomly distributed across the network, and robots (analogous to ants) are also randomly placed within this network. Each robot moves in a random trajectory and follows these rules:

a) If the robot does not carry an object, it decides at each node whether to pick up an object or move to the next node.
b) If the robot is already carrying an object, it decides whether to place it at the current node or move to the next node.

The decisions are probabilistic and depend on the local density of similar objects in the vicinity. Specifically:

a) If there are fewer objects of the same type nearby, the probability of taking that object is higher.
b) If there are more objects of the same type nearby, the probability of placing an object is greater.

In real-life ant behavior, similar objects may be identified through chemical or visual cues. For simplicity, the robots rely on the memory of the last 10 nodes they visited, which are considered the neighboring nodes for decision-making [9].

3.4.2 Clustering and Sorting Process

Initially, as the robots move according to the algorithm, small clusters of objects begin to form. Over time, as the probability of encountering similar objects near these clusters increases, further objects are deposited into the clusters, causing them to grow. This process is analogous to the grain growth mechanism during the solidification of metals, where each cluster acts as a nucleation site for growth.

When this experiment was conducted, after approximately 500,000 iterations, the objects were sorted into large, distinct clusters [10].

3.5 Search Algorithms

As mentioned in the previous algorithm, certain tasks can only be completed through the cooperation of multiple agents, while others can be accomplished individually but are more efficient when performed by a group. Sorting is one such task observed in ant colonies. In these colonies, brood and food are not uniformly distributed but are arranged in piles next to each other. Unlike humans, who rely heavily on mutual communication for successful sorting tasks, ants achieve this without any known mutual communication mechanisms or hierarchical organization. No single ant dictates decisions; instead, simple individual behaviors lead to the desired collective outcome.

Inspired by this behavior, a simple algorithm is proposed for robotic sorting. This algorithm requires only local environmental information and short-term individual memory. It is probabilistic and was tested using a Monte Carlo simulation to explain its functionality [11].

3.5.1 Ant Walk Algorithm

In technical terms, the task is to enable a group of ants to traverse all the paths of a directed graph without relying on internal memory. According to the Ant Walk algorithm, ants do not retain individual memory. Instead, when an ant visits a vertex, it examines all the edges connected to that vertex and moves along the edge with the lowest pheromone concentration [12]. After traversing a path, the ant deposits a random amount of pheromone. Pheromone concentration on each edge decreases with time, so the concentration at any given moment indicates how recently the edge was last visited. A lower concentration of pheromones means that a longer time has passed since the edge was last traversed. If multiple edges have the same lowest pheromone concentration, the ant makes a heuristic decision to choose its path. There are modifications of this algorithm inspired by the depth-first search algorithm and some in which the vertices are visited instead of the edges [13].

3.5.2 Hidden Markov Model

The Hidden Markov Model (HMM) is a model used to identify hidden patterns. One important problem associated with HMM is the estimation of the target parameters, as it is a machine learning approach. Several approaches, like Baum–Welch, Viterbi-Training, and Tabu-Search, have been used to solve the problem. However, each approach has its issues and drawbacks, for which ACO is a good alternative. Ants use pheromones for communication and to find the shortest path to solve a problem, like in the TSP [14]. The three main problems associated with HMM are: path finding, known parameter estimation, and unknown parameter estimation. Ants find the shortest path using the amount of pheromone present on the path. The algorithm described solves the problem by:

a) Finding the starting point by letting ants move in a random fashion
b) Set the initial pheromone
c) Estimates the weight of each edge and the vectors of each node
d) Make use of the pheromone to find the next path
e) Update the pheromone value based on the evaporation rate
f) Checks for the termination condition, and if found to be satisfied, then terminate
g) Or go to Step 3.

Finding the Initial Path

For the initial path, we observe a random ant moving in a graph from the source to the sink. The initial weights of the edges and the vectors of the nodes are taken to be zero.

Initial Pheromone

Two types of deposits of pheromones will be estimated in case of each ant, where the first one will be associated with the edges and the second one with the nodes. The quantity of pheromone present in an edge or a node indicates the number of ants traversed through that path. The greater the pheromone concentration, the greater will be the number of ants passed [15].

Estimation of Weights

In each iteration, the pheromone concentration is calculated using the residual amount, and then the weights and vectors of edges and nodes are normalized.

Finding New Path

The new path is found based on the weights and vectors of edges and nodes, respectively. The ant from a sequence S at node m_i in the first step of its movement, and the next sign c_k and state m_j s chooses as the path if:

$$j = \arg\max_{d} \frac{\left(log(w_{id}) + \tau_{id}^{A} \times \left(log\left(z_d(k)\right) + \tau_d^{E}(k)\right)\right)}{\sum_{r=1}^{n}\left(log(w_{ir}) + \tau_{ir}^{A} \times \left(log\left(z_r(k)\right) + \tau_r^{E}(k)\right)\right)} \tag{3.9}$$

Pheromone Evaporation

The weights are updated in every iteration based on the evaporation rate, and the pheromone concentration is recalculated.

Termination

The algorithm continues the iteration until

1. The number of iterations reaches a maximum value that was already determined.

   or

2. The algorithm converges [19].

3.6 Link Prediction Algorithm

The recording of many real-world data could have some data missing or incorrectly interpreted data. To resolve this, it is important to establish a link between them. These links could be temporary enough to predict future relations or to predict the missing data. Real-world systems could be taken as a network with the individual or the organization representing the nodes and the relationships between them as edges [16]. But the relationship or the link between these real-time networks is difficult to predict.

The link prediction algorithms can be broadly classified into three:

- *Generating Mechanism,* which is based on the local community paradigm, has comparatively low computational complexity.
- *Probabilistic Model* has a higher computational capability and is used for large-scale networks.
- *Model-free* has comparatively low computational power.

A quantum-based optimization algorithm is proposed to predict the missing links.

- A general assumption is made that a biased optimization algorithm is proposed to fix the incomplete dataset. This is based on the pheromones of an artificial ant that can show the importance of a path.
- Quantum bits and quantum logic gates are used in ACO [25].

3.6.1 Quantum-Based Optimization Algorithm

All the artificial ants are made to walk in a network randomly, with each link in the network having some amount of pheromones and visibility, which indicates the link between the nodes, where the number of ants passing through each link will be directly proportional to the pheromone concentration in the path. Once the ant traverses through the edge, the amount of pheromone will be updated [17]. According to quantum computation, the probability of the occurrence of each state varies concerning time. Likewise, the occurrence of a particular path depends on the pheromone density, which varies in accordance with time. The pheromone density determines the prediction.

- A random number of ants are made to travel a network with each link randomly assigned with a pheromone concentration and visibility.
- After each traverse, the pheromone density and the visibility are updated.

Here, we consider a network in which none of the links are interconnected. This algorithm assigns a score to each link based on visibility, which shows the probability of the link having a connection. The links with the highest score are likely to have a connection [18].

3.7 Application of Ant-Inspired Algorithms

3.7.1 Internet of Drones

Coverage path planning (CPP) is used to determine a path for the robot to cover all the points in a particular environment. In addition to path planning, CPP also includes mapping to avoid all obstacles in the traverse and environmental monitoring and inspection. Various bio-inspired solutions for the CPP have been derived from the ACO. Drones are widely used for inspection and monitoring purposes. They form a network, an internet of drones (IoD), to facilitate communication between multiple drones and perform multiple tasks simultaneously [19]. The IoD consists of a decentralized system where drones communicate with each other and a ground control station through wireless connections, often utilizing cellular networks.

Coverage path planning typically involves the region of concern being divided into several divisions, which could be based on some scenario or other constraints. An example of the division can be described as dividing the entire scenario into two. The first one includes cases where the bot needs to move only along the central line for purposes such as patrolling and surveillance. In the second scenario, the bot needs to scan the entire region, such as sweeping activity.

3.7.1.1 Problem Definition

The IoD is used to coordinate among different drones to avoid collision and congestion for the purpose of monitoring and inspection. It enables to traversal of multiple drones simultaneously. The problem described is to build an ACO-inspired system that could develop an IoD in an energy-efficient way.

- In the case of the IoD, only air constraints are taken into account. The division for the drones can also be based on road constraints.
- The model described here minimizes the energy consumed for the vertical and horizontal movement.

3.7.1.2 Key Aspects of the Algorithm

A drone inspired by an ant is typically called Anti-IoD. Anti-IoD starts with a drone, say, k, randomly selecting a node and an edge, traveling across it, making sure that all nodes and edges have been traversed. The algorithm also checks for any adjacent edges that have not been traversed. The non-traversed edges are added to the set of *possibe Edges* for a particular drone k. The drone k is the representation of an ant k.

The algorithm checks if any such edges are left in the set. And if no such adjacent edges are left that are not traversed yet, then the Breadth First search is used to find the path that is at the same distance from the current position as that of the last node. This selection of paths will result in a new table consisting of the paths in the order starting from the last traversed to the new. This is the table that obtained the best effective solution when compared to the previous table; then the current table is selected [20–22].

When the current node has no edges that remain untraversed, then it uses Breadth First to find the new node and assumes all the edges to be at an equal distance from the current node. It then makes use of the pheromone table to discover the next edge. The probability of taking the yth vertex, given the current position is the xth vertex with z^{th} vertex being already traversed is given by:

$$P_{xy}^{k} = \frac{[\tau_{xy}(t)]^{\alpha}[\eta_{xy}(t)]^{\beta}}{\sum_{l \in N_x^k}[\tau_{xl}(t)]^{\alpha}[\eta_{xl}(t)]^{\beta}} \tag{3.10}$$

where $\tau_{xy}(t)$ the pheromone concentration and $eta_{xy}(t)$ is the pheromone value.

The value of the pheromone changes every time the path is traversed using:

$$\tau_{xy}^{k}(t) = (1-\sigma)\tau_{xy}^{k}(t-1) + \sum_{k=1}^{K}\Delta\tau_{xy}^{k}(t) \tag{3.11}$$

where t is the number of traversals and $Delta\tau_{xy}^{k}(t)$ indicates the amount of pheromone deposited at each edge

$$\Delta\tau_{xy}^{k}(t) = \begin{cases} \frac{1}{E_k}, & (z,x,y) \in p_k \\ 0, & \text{otherwise} \end{cases} \tag{3.12}$$

E_k indicates the energy required for the drone k.

The energy required by the drone in the upward traverse is given by

$$E_{\text{up}} = 315d_1 - 211.261 \tag{3.13}$$

And the energy consumed in its downward motion is given by

$$E_{\text{down}} = 68.956d_2 - 65.183 \tag{3.14}$$

Here d_1 indicates the distance traveled in the vertical direction and d_2 indicates the distance traveled in the downward direction.

The algorithm terminates when the best path with the least energy consumed is returned. The experimental data provided the fact that the algorithm gives a better energy consumption [28].

3.8 Dynamic Task Allocation Problem – Machine Scheduling

In insect colonies, particularly ant colonies, there exist individuals with different specializations, similar to human societies. When a task arises within the colony, an individual decides whether to take it up based on their specialization. For example, in ants, there are two primary categories: *major ants* and *minor ants*.

Major ants typically perform critical functions, such as serving as soldiers or foragers, while minor ants handle tasks like collecting food from identified sources and organizing items within the nest. These roles are also influenced by the age of the ants: older ants often remain inside the colony to care for the younger ones, while adolescent ants venture out for foraging [23]. Although these roles seem rigid, experiments have shown that if the population of minor ants decreases for any reason, major ants can assume tasks usually performed by minors, and vice versa.

3.8.1 Problem Definition

A dynamic task allocation problem is a type of machine scheduling problem in which agents or machines have varying specializations. Traditional scheduling algorithms often fail to account for this variability. Inspired by insect behavior, a new algorithm is proposed to address this problem.

The specific problem considered involves trucks leaving an assembly line that must be assigned to painting booths. Here:

- Trucks represent *jobs*, each associated with a specific color to be painted.
- Painting booths represent *machines*, each with distinct capabilities.
- Each booth has a fixed queue length.

For each job type (e.g., a specific color), every machine has an associated *threshold value*, which is dynamically updated. Similarly, each job has a *stimulus value*. When a new truck (job) arrives, a machine decides whether to accept it based on a probabilistic distribution determined by the machine's threshold value and the job's stimulus value.

3.8.2 Key Aspects of the Algorithm

Each machine (or agent) has a specific threshold value associated with each task type. This threshold value represents how reluctant a machine is to accept a certain task. Machines with higher threshold values for a given task are less likely to accept that task, whereas lower threshold values indicate a higher likelihood of acceptance. On the other hand, each task has a stimulus value, which reflects the task's urgency or appeal. A higher stimulus value increases the probability that a machine will take up the task [24]. When a new task arrives, the machine decides whether to accept it based on the relationship between its threshold value and the task's stimulus value. The decision is made probabilistically, meaning there is no strict rule but rather a likelihood of acceptance influenced by these two factors. Once a machine accepts a task, its threshold value for that specific task is updated. This update reflects that the machine becomes more experienced or specialized in handling that task [25]. Consequently, the machine becomes slightly more inclined to accept similar tasks in the future.

Additionally, the algorithm considers setup time when switching tasks. For example, if a machine is currently handling one task type and needs to switch to another, the associated setup time is taken into account. To discourage frequent switching, the threshold value for the new task is temporarily adjusted to make it less likely to be chosen unless necessary. Finally, when one machine takes up a task, the algorithm increases the threshold values for other machines for that specific task [26]. This adjustment reduces the likelihood of the same task being accepted by other machines, balancing the workload and ensuring specialization. Equations used

The probability of booth k to get engaged in task j is given by:

$$P(s_{c_j},\theta_{k,c_j}) = \frac{s_{c_j}^2}{s_{c_j}^2 + \alpha\theta_{k,c_j}^2 + \Delta T_k^2 \beta(j)} \tag{3.15}$$

where:

- s_{c_j}: Stimulus associated with the truck j for color c_j.
- θ_{k,c_j}: Threshold of agent k for color c_j.
- α, β: System parameters.
- $\Delta T_k(j)$: Time variable

When truck j is assigned to booth k, the threshold values are updated as follows:

$$\theta_{k,c_j} \to \theta_{k,c_j} - \xi \tag{3.16}$$

$$\theta_{m,c_j} \to \theta_{m,c_j} + \phi, \quad \forall m \neq k \;\; [26] \tag{3.17}$$

where:

- ξ: Learning factor that decreases the threshold for booth k.
- ϕ: Forgetting factor that increases the threshold for all other booths m for the same color c_j.
- θ_{m,c_j}: Threshold of other booths m for color c_j.

Time variable ($\Delta T_k(j)$) is the time until task j starts to be painted in booth k, and is determined by the following equation:

$$\Delta T_k(j) = qt^{proc} + nt^{setup} + t^{working} \tag{3.18}$$

where:

- q: Number of trucks in the queue of booth k.
- t^{proc}: Time required to paint one truck.
- n: Number of setups required for the trucks in the booth's queue.
- t^{setup}: Time required for a setup.
- $t^{working}$: Time necessary to finish the currently painted truck [26].

3.9 Additive Manufacturing Tool Path Optimization Using Ant Colony Optimization

Additive manufacturing, commonly known as 3D printing, is a rapidly evolving technology. While tool paths for conventional machining processes are well-optimized, the tool paths for additive manufacturing remain suboptimal. To achieve optimal mechanical properties in 3D-printed parts, the part is typically divided into multiple segments, and an infill pattern is applied within these segments [27]. Additionally, a single 3D print layer may contain multiple disjoint parts, all of which must be covered during printing.

A significant challenge in extrusion-based 3D printing is the excessive time wasted during nozzle transitions—movements between different sections or parts. These transitions not only increase printing time but also lead to the formation of strings, which degrade surface quality and necessitate additional post-processing effort. To address these issues, a multi-layered ACO algorithm was proposed by Fok *et al.* The problem is modeled as an undirected rural postman problem, where each node (representing a part or segment) must be visited at least once. This optimization is performed in two layers:

1. At the first layer, nodes correspond to disjoint parts within a single layer.
2. At the second layer, nodes represent individual segments within each part.

Movements within a segment are not included in the objective function, as they are governed by predefined infill patterns dictated by material property requirements. The objective function focuses on minimizing the time taken for nozzle transitions between sections and the retraction time required for such movements. Transition times are calculated based on the nozzle's acceleration and deceleration profiles as specified by the machine, and these are used to create a time-cost function minimized by the algorithm [28].

3.9.1 Methodology

The algorithm follows these steps:

a. *Input Model:* A sliced 3D model is provided, and all print segments for each layer are identified.
b. *First Layer of Optimization:* For layers containing multiple disjoint parts, the algorithm determines the optimal sequence for traversing these parts to minimize transition time.
c. *Second Layer of Optimization:* Within each part, the algorithm optimizes the traversal sequence of individual segments.
d. *Full Model Optimization:* The process is repeated across all layers to obtain the optimized path for the entire print.

3.10 Conclusions

Ant Colony Optimization-inspired solutions, based on the collective nature of ant colonies, are used for solving various problems. The foraging behavior of ants helps find the shortest path with the help of the pheromone trails they leave behind, and the rate of evaporation. Various algorithms discussed above, like the MAX-MIN ant system, Ant System, AddACO, and its variants, make use of this trait of ants to come up with a solution for the salesman problem. Apart from these, various other traits were also identified, including the swarm algorithm, which enables communication among various robots, and were also discovered and discussed above. Different algorithms were discussed, which could be used to identify the hidden patterns and predict the links between the nodes and edges they travel.

Every algorithm involves identifying the amount of pheromones left behind and the ability of the ants to trail based on the evaporation rate. The shortest and most energy-efficient path would have a higher concentration of pheromones. The prediction of the links between the nodes and edges is also based on the concentration of pheromones and their visibility. Various other applications were also discussed above, which involve the IoD, optimum job allocation for machines, and path optimization for advanced machining processes. Ant Colony Optimization proposes better and more time-efficient solutions for problems that are difficult for mankind to solve.

3.11 Future Research Opportunities

Even though ACO-inspired solutions provide better solutions for various problems, there are still certain instabilities and inefficiencies. The swarm solution fails when too many robots need to be controlled simultaneously, as in the case of the IoD, which requires optimization. Additionally, there are various fields where ACO-inspired solution integration could be utilized, such as integrating with artificial intelligence, machine learning, and real-time optimization. Energy consumption efficiency is one such area that requires further development. These factors make ACO a promising field for research.

References

1. Dorigo, M. and Stützle, T., Ant colony optimization: Overview and recent advances. *Swarm Intell.*, 1, 1, 1–29, 2004.
2. Bonabeau, E., Dorigo, M., Theraulaz, G., Inspiration for optimization from social insect behaviour. *Nature*, 406, 39–42, 2000.
3. Di Caro, G. and Dorigo, M., AntNet: Distributed stigmergetic control for communications networks. *J. Artif. Intell. Res.*, 9, 317–365, 1998.
4. Khan, F.A., Ullah, K., Rahman, A.U., Anwar, S., Energy optimization in smart urban buildings using bio-inspired ant colony optimization. *Soft Comput.*, 27, 973–989, 2023.
5. Jabbar, S., Iram, R., Minhas, A.A., Shafi, I., Khalid, S., Ahmad, M., Intelligent optimization of wireless sensor networks through bio-inspired computing: Survey and future directions. *Sci. World J.*, 421084, 2013.
6. Dorigo, M., Maniezzo, V., Colorni, A., Ant system: Optimization by a colony of cooperating agents. *IEEE Trans. Syst. Man Cybern. B Cybern.*, 26, 1, 29–41, 1996.
7. Dorigo, M. and Gambardella, L.M., Ant colony system: A cooperative learning approach to the traveling salesman problem. *IEEE Trans. Evol. Comput.*, 1, 1, 53–66, 1997.
8. Socha, K. and Dorigo, M., Ant colony optimization for continuous domains. *Eur. J. Oper. Res.*, 185, 3, 1155–1173, 2008.
9. Gambardella, L.M. and Dorigo, M., Solving symmetric and asymmetric TSPs by ant colonies, in: *Proc. IEEE Int. Conf. Evol. Comput*, pp. 622–627, 1996.
10. Gutjahr, W.J., A graph-based ant system and its convergence. *Future Gener. Comput. Syst.*, 16, 8, 873–888, 2000.
11. Bullnheimer, B., Hartl, R.F., Strauss, C., A new rank-based version of the ant system: A computational study. *Cent. Eur. J. Oper. Res.*, 7, 1, 25–38, 1999.
12. Maniezzo, V., Exact and approximate nondeterministic tree-search procedures for the quadratic assignment problem. *INFORMS J. Comput.*, 11, 4, 358–369, 1999.
13. Stützle, T. and Hoos, H.H., MAX–MIN ant system. *Future Gener. Comput. Syst.*, 16, 8, 889–914, 2000.
14. Blum, C., Ant colony optimization: Introduction and recent trends. *Phys. Life Rev.*, 2, 4, 353–373, 2005.

15. Dorigo, M. and Birattari, M., Ant colony optimization, in: *Encyclopedia of Machine Learning*, pp. 36–39, 2011.
16. Kumar, A. and Singh, M., A survey on ant colony optimization with emphasis on its application to routing in mobile ad hoc networks. *Artif. Intell. Rev.*, 43, 1, 27–63, 2015.
17. Zhou, Y., Wang, L., Wang, Y., Wang, Y., An improved ant colony optimization algorithm for solving the traveling salesman problem. *Appl. Soft Comput.*, 37, 643–653, 2016.
18. Dorigo, M., Birattari, M., Stützle, T., Ant colony optimization. *IEEE Comput. Intell. Mag.*, 1, 4, 28–39, Nov. 2006. doi: 10.1109/MCI.2006.329691.
19. Scianna, M., The AddACO: A bio-inspired modified version of the ant colony optimization algorithm to solve travel salesman problems. *Math. Comput. Simul*, 218, 357–382, Apr. 2024. doi: 10.1016/j.matcom.2023.12.003.
20. Eslami, N., Yazdani, S., Mirzaei, M., Hadavandi, E., Aphid–Ant Mutualism: A novel nature-inspired metaheuristic algorithm for solving optimization problems. *Math. Comput. Simul*, 201, 362–395, Nov. 2022. doi: 10.1016/j.matcom.2022.05.015.
21. Kube, C.R. and Zhang, H., Collective Robotics: From Social Insects to Robots, 1993.
22. Deneubourg, J.-L., *et al.*, The dynamics of collective sorting robot-like ants and ant-like robots. *J. Theor. Biol.*, 143, 2, 259–274, 1990. [Online]. Available: https://www.researchgate.net/publication/235362107.
23. Bruckstein, A.M., Wagner, I.A., Lindenbaum, M., Smell as a computational resource: A lesson we can learn from the ant. *Informatics*, 23, 4, 247–256, 1996. [Online]. Available: https://www.researchgate.net/publication/221342108.
24. Emdadi, A., Ahmadi Moughari, F., Yassaee Meybodi, F., Eslahchi, C., A novel algorithm for parameter estimation of Hidden Markov Model inspired by Ant Colony Optimization. *Heliyon*, 5, 3, e01299, Mar. 2019. doi: 10.1016/j.heliyon.2019.e01299.
25. Cao, Z., Zhang, Y., Guan, J., Zhou, S., Link prediction based on quantum-inspired Ant Colony Optimization. *Sci. Rep.*, 8, 1, 31254, Dec. 2018. doi: 10.1038/s41598-018-31254-3.
26. Bine, L.M.S., Boukerche, A., Ruiz, L.B., Loureiro, A.A.F., A novel Ant Colony-inspired coverage path planning for Internet of

Drones. *Comput. Netw.*, 235, 109963, Nov. 2023. doi: 10.1016/j.comnet.2023.109963.

27. Birattari, M. and Dorigo, M., An insect-based algorithm for the dynamic task allocation problem, 2005. [Online]. Available: https://www.researchgate.net/publication/220633907.
28. Fok, K.Y., Cheng, C.T., Ganganath, N., Iu, H.H.C., Tse, C.K., An ACO-based tool-path optimizer for 3-D printing applications. *IEEE Trans. Ind. Inform.*, 15, 4, 2277–2287, Apr. 2019. doi: 10.1109/TII.2018.2889740.

4

Bio-Inspired Fabrication for Sustainable Water Scarcity Solutions: Learning from Nature

Abstract

One of the most common problems the world is facing is water scarcity. Despite many advanced technologies and advanced manufacturing techniques, it is impossible to overcome the global problem of water scarcity. Researchers are investing in bio-inspired water collection and preservation systems. To study deeper aspects of nature, the Namib Desert provides them with an outstanding water harvesting model system. This chapter deals with beetles' remarkable ability to collect water from nature, and their hydrophobic or hydrophilic portions make them exceptionally good water harvesters. It is also interesting to study the transportation mechanism. To have a deep understanding, mechanical structure design and its dynamics must be considered, along with surface and subsurface characteristics. Advanced manufacturing techniques are available for the fabrication of such a structure with good dimensional and geometrical accuracy. The mechanism of oil and water separation, along with the biosensing mechanism, makes it more challenging to produce an anti-icing subsurface. This chapter also presents a critical analysis of the application of technologies in bio-sensing and fog harvesting to get sustainable and environmentally friendly water harvesting solutions to solve global problems.

***Keywords*:** Namib Desert beetle, biomimicry, hydrophilic, hydrophobic, water harvesting

Jagadeesha T. and Sandip Kunar. Bio-Inspired Manufacturing Systems, (85–108) © 2026 Scrivener Publishing LLC

4.1 Introduction

Water scarcity poses a significant threat to human survival, exacerbated by climate change, population growth, and increasing demand for freshwater. According to the United Nations, nearly two billion people reside in regions facing severe water stress, and by 2025, half of the world's population may experience water shortages [1]. This dire scenario necessitates the development of innovative and sustainable water collection methods. Water is a necessity for human civilization to exist and survive. Without water, there is no life on Earth. High population, technological advancement, industrialization, deforestation, and soil erosion created water scarcity on Earth. It is said that the next war would be over water. There are more than 2 billion people on Earth who do not have access to clean water [2]. There are many technological advancements in retrieving water from nature. Some of them are extracting water from fog or humidity. There are many coastal places in India where the humidity is quite high; in such regions, water from atmospheric humidity is a viable choice for water harvesting [3]. But there are many drawbacks, such as very low efficiency and suitability. Conventionally, researchers used mesh nets and coatings to harvest water [4]. Researchers began to search for alternative technologies by studying nature, and many bio-inspired designs are possible for water harvesting. One of the fascinating creatures is a beetle. This chapter presents a water harvesting method that beetles have adapted and how bio-inspired design based on beetles can mitigate the water scarcity problem [5].

4.2 Inspiration from Beetles

The deep insight into beetle water harvesting is not only surprising but also worth copying. Beetles collect the moisture from the surrounding fag using their own exoskeletons. Two regions exist on a beetle. One surface is hydrophobic, and the other is hydrophilic. The hydrophilic surface ensures that water extracted from fog sticks to the beetle, while the hydrophobic surface ensures that water droplets go toward the mouth for consumption [6]. Close microscopic examination of the exoskeleton using sophisticated characterization techniques reveals that the shell is made of micro- and nanostructures. These micro- and nanostructures increase the water-retaining ability, and this facilitates the condensation at a faster rate. Both micro- and nanostructures can be created using laser techniques on structures. Researchers also developed new critical subsurfaces that can hold and transport water [7]. Chemical etching and lithography are other techniques using which micro and nano structures can be created, and deposition can be done using chemical or physical vapor deposition techniques [8].

The surfaces are associated with surface energies, and beetles have not only micro- and nano-structured surfaces but also a gradient from micro to nano structures and scales. These gradients make it more appropriate for biosensing and condensation of the droplets. The angle of attachment and detachment, depending upon the hydrophilicity and hydrophobicity, makes it more water harvesting efficient [9].

Table 4.1 Different beetles, bio-inspiration, and application.

Types of Beetles	Bio inspiration	Application	Device/product
Namib Desert Beetle	Hydrophilic bumps and grooves	Fog and dew harvesting	Fog nets, coating
Stenocara Beetle	Graded hydrophobic and hydrophilic surfaces	Droplet collection	Polymer coating for water collection
Cybister Beetle	Micro- and nanostructures	Water repellent	Self-cleaning textiles and paints
Tenebrionid Beetle	Dorsal structure	Passive water collection system	Building and air-conditioning units
Darkling Beetles	Directional droplet flow	Water drainage system	Roof coating

Due to advancements in manufacturing processes, as well as characterization processes, it is possible to create such a graded structure using lasers, electron beam, electrospinning, lithography, and etching [10, 11]. Beetles not only have functionally gradient structures but also have the ability to separate water from oil and anti-icing abilities. This enables beetles to survive in a very harsh environment. Various beetles provided different inspiration to researchers, leading them to develop various products and devices, as shown in Table 4.1.

4.3 Water Harvesting Process

This chapter presents the mechanism of water harvesting by beetles concisely by covering the following important topics: investigating the potential of beetle-inspired water harvesting systems to address water scarcity and related challenges. Specifically, it aims to:

- How beetles collect water in harsh environments like deserts.
- Fabrication techniques that are available for reproducing gradient surfaces.
- Modeling, manufacturing, and simulation challenges from a practical point of view.
- How these are realized for industrial and commercial uses for fog harvesting and biosensing.

By proper analysis of beetle structures, the water harvesting mechanism can be understood and mimicked to solve water scarcity issues, which is a global problem.

4.4 Fabrication of Beetle-Inspired Surfaces

4.4.1 Overview of Fabrication Techniques

The advances in semiconductor manufacturing contributed a lot to producing micro- and nanostructures. Advances in lithography, thin film deposition, etching, and spinning made it possible to produce nano-level electronics structures such as CMOS devices. Artificial surfaces and their gradient can be created using these technologies, and advances in manufacturing systems allow us to optimize the process parameters, machine parameters, and environmental parameters to produce the surfaces and devices more effectively within a short product life cycle time. A clean room used for lithography makes it possible to etch micro- and nano-level structures, and etching technology can produce very fine structures with high aspect ratios [12].

Electrospinning

Superior texturing is observed on beetles' surfaces, and many researchers have tried to use conventional electrospinning techniques to reproduce these textured surfaces. Textured surfaces can be produced by many processes using beam processes, such as lasers. In electrospinning, very fine fibers are produced using the electrospinning process. After producing such fine nano fibers, they are deposited on a base or substrate. Details are shown in Figure 4.1. By proper spinning and alignment, it is possible to create super fine meshes that resemble the textured surface on beetles. Roughness and resolution are important factors because hydrophobicity depends on roughness and gap. Many polymers are used to make nanostructures,

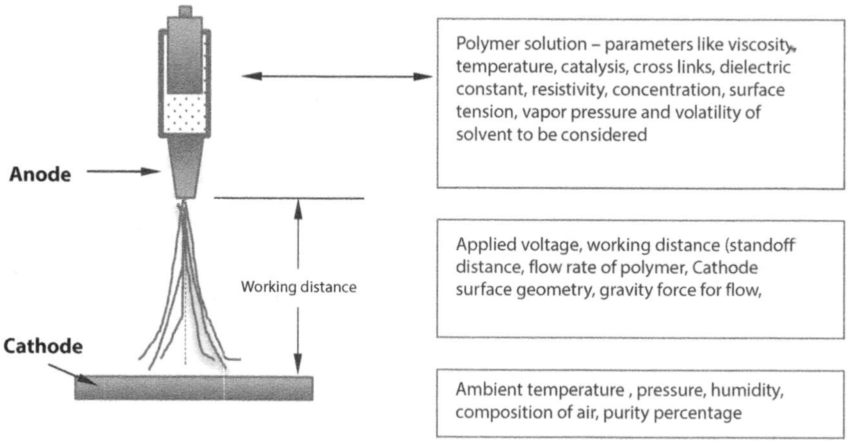

Figure 4.1 Electrospinning processes.

but polyacrylonitrile is most commonly used. Producing a lab-scale specimen by electrospinning is easy, but producing a mass production rate remains a challenge [13].

Inkjet Printing

Inkjet printing is one of the common and old technologies available for thin film deposition. Complex patterned surfaces can easily be printed using inkjet technology. Very precise control is possible, which can create a gradient-statured material at the nano- and microscale and hydrophilic/hydrophobic surfaces. This is a cost-effective process for producing structures with high levels of precision and repeatability. Here, ink itself is used to produce gradient surfaces and subsurface [14]. A representative diagram is shown in Figure 4.2. Unlike the electrospinning process, this inkjet printing can be used to produce parts in mass production with an acceptable amount of product variation. But there are practical issues regarding the longevity of printed materials under the exposure of UV light, weather conditions, etc.

Laser Etching

Laser properties, such as monochromaticity, exceptional coherence, both temporal and spatial coherence, high intensity, high power, and high resolution, can be used to produce nano-level structures on the substrate. Laser etching processes are shown in Figure 4.3. Highly

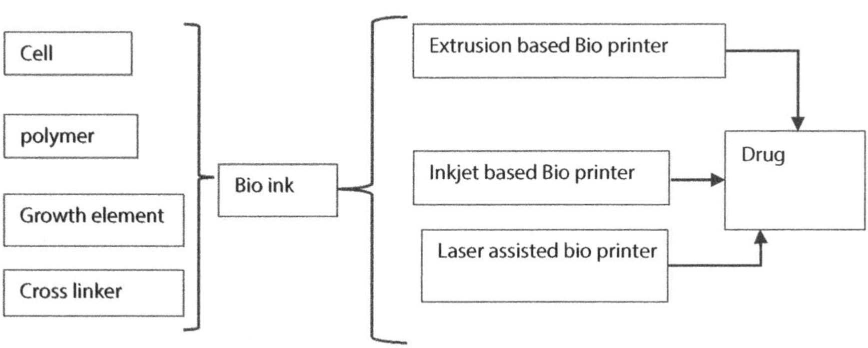

Figure 4.2 Inkjet processes.

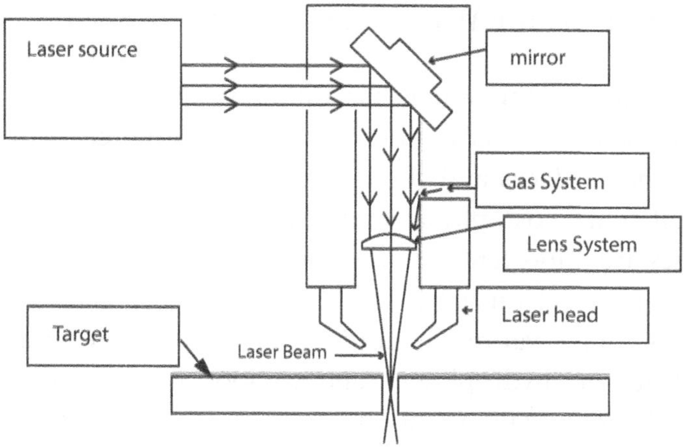

Figure 4.3 Laser processes.

precise patterns are possible with laser etching. Process parameters can be adjusted easily to mimic the exoskeleton pattern of a beetle. Optimized process and machine parameters can be used to produce a surface of desired wettability for water harvesting and biosensing. But the high cost of equipment and maintenance limits its use for mass production of textured surfaces for water harvesting, anti-icing, and biosensing applications [15].

Photolithography

In the semiconductor or wafer fabrication industry, high-resolution advanced photolithography techniques are used to produce a patterned surface for electronic devices. The same technology can be used to create a pattern surface for water harvesting. Exceptional precision, resolution, and focus made it adoptable and acceptable for the production of micro- and nano-gradient surfaces on substrates or wafers. Usually, a substrate is coated with photoresist material, and a pattern is created using photolithography since photoresist is sensitive to UV light exposure. Then, etching is used to create the pockets or patterns. It is fast and dependable, but it is expensive because mask creation is time-consuming and expensive [16].

4.5 Materials for Water Harvesting

The efficiency of the water harvesting systems depends on the material used. Materials must be able to produce hydrophilic/hydrophobic surfaces and should exhibit good wettability properties. Availability, cost, physical, mechanical, electrical, and thermal properties are important for material selection for water harvesting [17]. The material must be durable, reliable, and have consistent properties under various conditions of temperature and humidity. Hydrophilic, hydrophobic, and composite materials are used. Hydrophilic materials are useful for condensing water droplets. Many polymers, along with a wide range of nanoparticles, are used to produce water-attracting surfaces. Prominent ones are polyvinyl alcohol and polyethylene glycol. Both are known for forming a stable nanostructure. Hydrophobic materials, such as polytetrafluoroethylene and polydimethylsiloxane, are commonly used since they have a high wettability angle. There are many hydrophobic coatings available to create a desired hydrophobic surface with good durability and good water harvesting properties [18]. Composite materials can be tailor-made to produce desired hydrophilic or hydrophobic surfaces. Desired direction-dependent properties can be achieved with the proper choice of reinforcement and matrix element.

4.6 Challenges

Several advanced technologies available today include wafer fabrication technology, yet there are manufacturing difficulties that are listed below.

1. The degradation of polymeric materials is a common problem. Due to its low glass transition temperature and poor mechanical properties, maintaining the pattern for a long time against UV exposure is very challenging. Another aspect is poor tribological

characteristics, which hinder its practical applications in water harvesting.

2. Many surfaces are easy to realize in the lab, but they are difficult to manufacture in large quantities and also in larger sizes. The properties of nano- and microstructures vary when they are scaled up from a manufacturing point of view. Currently available techniques need further technological refinement for the large-scale manufacturing of surfaces suitable for water harvesting.
3. Many advanced techniques like laser etching and photolithography are expensive, and they prohibit the production in large no. They are not cost-effective if they are produced in batches of 5 or 10. Mask cost in photolithography can only be compensated if they are produced in huge quantities, such as electronic chips.

4.7 Mechanism of Water Harvesting

Micro- and nano-graded structures, materials, and wettability are important parameters that govern the efficiency of water harvesting systems. There are three stages: capturing water, condensation of water captured, and a transportation mechanism to get the harvested water.

4.7.1 Interaction Between Hydrophilic-Hydrophobic Surfaces

Hydrophilic surfaces attract moisture present in the atmosphere, and then it condenses. There is growth of the droplets, and finally, bigger water droplets are captured. This can be modeled using a Young-Laplace equation, which is given by:

The effectiveness of this behavior is governed by the Young-Laplace equation:

$$Pressure\ difference = \Delta P = \frac{2\gamma}{r}$$

ΔP is the pressure difference inside and outside the droplet.
γ is the surface tension of water.
r is the radius of curvature of the droplet.

It is evident from the equation that if the pressure difference is large, then there is a greater possibility of condensation and an increase in growth size.

Hydrophobic surfaces repel water, and it is very much needed to detach water droplets from the surface; otherwise, transportation of the collected water through condensation becomes difficult. Surface energy gradient is essential for achieving the effectiveness of repulsion and collection of water droplets in water harvesting systems.

In water harvesting, two contradicting surfaces are needed. Hydrophobic surfaces are needed to attach water particles; they are also needed to detach the particles and transport them. Transportation efficiency depends on the wettability of the surfaces and sub-surfaces. Strategic placement of these two surfaces is quite important, and the gradient surface energy plays an important role in achieving good transportation efficiency.

4.7.2 Wetting Models

Many researchers modeled wetting effectively and cross-verified with experimental results. Two prominent models used in the analysis of surfaces are the Wenzel model and the Cassie–Baxter model.

4.7.2.1 Wenzel Model

This model considers surface roughness as an important parameter in controlling the wettability of surfaces. Adhesion is increased by increasing the surface roughness. Adhesion is expressed in terms of the apparent contact angle $\varnothing_p$

The apparent contact angle (θ_w) is given by:

$$cos(\varnothing_p) = rcos(\theta_i)$$

r gives the ratio of actual surface area to actual flat area.
θ_i is the intrinsic contact angle on a smooth surface

4.7.2.2 *Cassie–Baxter Model*

If surfaces have roughness with a good number of valleys on the surface, then the contact angle differs compared to a flat surface. If more valleys are present, then a droplet of liquid sits on the peak of textured, graded structures. To calculate the apparent contact angle, the Cassie–Baxter model is used.

$$cos(\theta_a) = s_1 cos(\theta_{ic1}) + s_2 cos(\theta_{ic2})$$

S_1 and S_2 represents the area covered by intrinsic contact angles θ_{ic1} and θ_{ic2}.

This model applies to hydrophobic areas because hydrophobic surfaces offer the least resistance for water droplet transportation.

4.7.2.3 *Dynamic Wetting*

Micro- and nano-structures on a beetle's exoskeleton offer both advantages of the Wenzel and Cassie–Baxter models. Hydrophilic surfaces are modeled by the Wenzel model, which is essential for water collection. Hydrophobic surfaces are modeled by the Cassie–Baxter model, which is essential for water detachment and transportation. Thus, the beetle combines both models and offers dynamic wetting [19].

4.8 Comparison with Other Natural Systems

Nature has created many other efficient water harvesting plants/insects/animals. Notable ones are cactus spines, spider silk, and pitcher plants. A cactus has beautiful spines, which are multifunctional.

Spines protect the cactus from its enemies, and they also provide shade. In addition to that, the tips of the cactus are hydrophilic. This helps in absorbing moisture, converting it into water, and transporting it to the plant base. Water flows due to gravitational force rather than surface energy differences.

Spider silk has a combination of hydrophilic and hydrophobic structures. Through a complex web, dews are collected from the circumference and transported toward the web's center. It is less efficient compared to the beetle in terms of fog harvesting, but good at dew collection. Efficient dew collection models are developed based on spider silk's behavior.

The pitcher plant has two distinct surfaces. Internal surfaces are hydrophilic, which collect the water, and external surfaces are hydrophobic, which transport water to the bottom. The water collection and transportation efficiency depend on the humidity of the surroundings. If the humidity is high, the efficiency of the pitcher plant in water harvesting is high, and vice versa. They are great models for rainwater harvesting rather than fog collection, bio-sensing, and anti-icing.

4.9 Mathematical Modeling and Analysis

Mathematical models are developed to understand the water harvesting mechanism, thereby making optimization possible through simulations. Models are available for droplet formation and transportation, surface energy and gradients, optimization, and overall systems improvements.

4.9.1 Formation of Droplet and Related Surface Energy

Hydrophilic surfaces attract moisture present in the atmosphere, and then it condenses. There is growth of the droplets, and finally, bigger water droplets are captured. This can be modeled using a Young-Laplace equation, which is given by:

$$Pressure\ difference = \Delta P = \frac{2\gamma}{r}$$

ΔP is the pressure difference inside and outside the droplet.
γ is the surface tension of water.
r is the radius of curvature of the droplet.

Curvature and energy associated with the surface are important in the analysis of droplet formation, nucleation, growth, detachment, and transportation. From the above equation, the pressure difference is very important in condensation and the growth of the droplets. It is surprising but true that beetles optimize and amplify condensation and growth by nanoscale roughness on their exoskeleton, and is mimicked by many researchers [19].

4.9.2 Dynamics of Growth of Drops and Collection

The growth of water droplets can be modeled to evaluate the collection efficiency of beetle-inspired surfaces. The volume of a growing droplet (V) is expressed as:

$$V_d = \frac{1}{2}\rho_w g\left(\frac{r_w}{\gamma}\right)^2 t^2$$

V_d = Total volume of water harvested
ρ_w = Density of water
g = Acceleration due to gravity
r = Radius of curvature of the droplet collected
γ = Surface tension
t = Collection time

It is clear from the above equation that water efficiency is directly related to surface tension and curvature of the droplet. If the surface is hydrophilic, the small droplets are naturally obtained; on the other hand, if the surface is hydrophobic, large droplets are obtained. Additionally, surface-texturing and gradients are important. Surface energy associated with hydrophobic and hydrophilic surfaces is a decisive factor in droplet water harvesting [20]. Adhesion and transportation play an important role in water harvesting. More adhesion implies less transportation and vice versa. If the contact area is larger, the possibility of rolling off the surface is smaller; if the contact area is smaller, then adhesion is less, and transportation of droplets becomes easy. Small droplets stick to hydrophilic surfaces, and if the textures are quite nearby, then small droplets join together and form a bigger droplet, depending upon the contact area and surface energy due to adhesion. Bigger droplets get transported to the required spot. Rolling resistance also depends on the thermal and rheological characteristics of water and the environment in which water is harvested. Both surfaces must interact to get good fog harvesting [21]. Surface texturing is important to grow the droplets. Various methods, like laser, electrospinning, and etching, are most commonly used to get the desired surface texture. The type of fabrication depends on the cost and the quantity to be made. It is quite challenging to find a manufacturing process to produce these on a larger scale.

The efficiency of this system is measured in terms of the growth rate of droplets and the total volume of water harvested. The design of surfaces, texturing, energy levels, gradients, and curvature are decisive factors in arriving at good efficiency of fog and water harvesting. A proper understanding of water-harvesting mechanisms helps in the design of proper surfaces for water harvesting.

4.9.3 Contact Angle and Surface Energy

The efficiency of water harvesting depends on two important parameters. One is energy-related, and the other is wettability-related. Energy-related properties are expressed in terms of surface energy, while wettability is measured and quantified using contact angle measurement.

Contact angle measurements allow the fluid to wet the surface; the wet angle is about the hydrophilic or hydrophobic nature of the fluid.

The surface properties on which the wettability angle is measured are also considered while evaluating the total functionality of the surface [22] between the advancing and receding contact angles of a water droplet. Low hysteresis ensures that droplets move smoothly across the surface. The contact angle hysteresis is calculated as:

$$Contact\ angle\ hystersis = \theta_A - \theta_R$$

θ_A = contact angle when the droplet is advancing.
θ_R = contact angle when the droplet is retracting.

Surface Energy Gradient: The smooth gradient between the wettability angle associated with hydrophilic and hydrophobic regions is very important in water harvesting. There are two important phenomena: one is the capturing of water droplets, and the other is the rolling off of the droplets. The smoothness of the gradient decides the efficiency of the water collection and delivery systems.

4.9.4 Performance Optimization Models

The characteristics, influencing parameters, and effects of the control parameters can be studied by mathematical modeling and numerical simulations. Now, custom-made and general commercial software is available with high-power computational ability with high speed. The interrelation and interdependence of various influencing factors, such as surface texture, material properties, and surrounding conditions, can be studied using mathematical models and simulations. Many researchers formulated droplet growth, roughness, water transport, and harvesting problems, which obtained good results [23].

Surface roughness and droplet growth rate are interrelated. If the textured surface has fewer resolutions created by lasers or electrospinning, then the combination of small water droplets to form a big droplet can be modelled using the droplet growth model, which can be expressed as:

$$R_G = \beta\left(\frac{A_{eff}}{\gamma}\right)\Delta t$$

R_G = growth rate of the droplet,
β = constant depending on the material used.
A_{eff} = effective surface area
γ = surface tension of water,
Δt = the time of condensation

Droplet Transport Efficiency Model: As discussed earlier, the transportation of water particles can be modeled using the Wenzel model and the Cassie–Baxter model. Transportation efficiency can be modelled using the transportation force.

$$F_{TR} = \frac{\varpi}{\varphi_R}$$

F_{TR} = transport force of the droplets,
ϖ = constant relating to the roughness of the surface,
φ_R = contact angle hysteresis.

Higher transportation efficiency is possible with lower contact angle hysteresis. Optimized surfaces can reduce the contact angle hysteresis. The optimization of all parameters is important, especially when devices are to be produced in large quantities.

4.9.5 Scalability Models

The beetle exoskeleton is small but has an excellent water-harvesting mechanism. However, in real practice, a small area produces very low yield; therefore, scale-up models are necessary from the manufacturing point of view. High-scale models are needed for high water collection rates.

Scaled-Up Water-Harvesting: Before scaling up the model to the required water-harvesting system, it is important to know the total surface area available and the total moisture in the surroundings. The total efficiency can be calculated using the following equation.

$$\eta_O = \eta_S \times A_e \times \Phi_h \times \eta$$

η_o = overall water collection efficiency,
η_s = efficiency of the available surface
A_e = effective surface available for water harvesting
Φ_h = humidity factors
η = efficiency factor

Environmental Conditions: The morphology of the surface is not the only thing that matters in harvesting the water; environmental conditions also play a very important role, as well, since the water has to be extracted from fog, dew, or moisture in the atmosphere, which depends on the psychometric properties of air, whether saturated. Condensation and droplet transpiration is a complex phenomenon and is expressed using

$$\eta_e = \eta_h \times \eta_t \left[exp\left(-\frac{v}{v_t} \right) \right]$$

η_e = environmental efficiency factor
η_h = moisture-related efficiency.
v = speed of the wind.
v_t = threshold wind speed
η_t = temperature-dependent efficiency factor.

4.10 Applications Based on Water-Harvesting Mechanisms of the Beetle

The unique and complex design of the beetle's water-harvesting mechanism inspired many researchers to design for water conservation. Some of them are listed below

1. Fog harvesting is done by capturing moisture from humid atmospheric air. This is done by adding nano- and microstructures and integrating gradient structures.
2. Anti-icing and condensation control mechanism. This mechanism is used to prevent ice formation in cold and harsh environments. Adhesion control is important for transportation. This is useful in aerospace applications, as the building of ice on wings is common and must be controlled from a safety point of view. Sensor technology can be used to sense, and active and passive controls are possible
3. The water oil separation mechanism of oil and water separation works well on graded micro- and nano-textured surfaces. The ability to select and separate the required component, like water, is done by the proper design of surfaces. Surfaces that attract water but let go of oil must be designed for effective water-oil separation.
4. Biosensing: Lab-on-chip is an integrated approach used to prevent downtime in the medical field, and biosensing can be incorporated on lab-on-chip with advanced electronics.
5. Microfluidics: a transportation mechanism used in textured gradient surfaces eliminates the need for pumping systems in microfluidics systems. This application can be found in portable handheld medical devices and the characterization of portable hand-held tools for chemicals, medicines, especially in remote areas. The reduction of water evaluation leads to a high yield of water collection, and condensation through proper design yields good water transportation, improving efficiency.

4.11 Conclusion

Water scarcity is one of the biggest global problems. There are many reasons for water scarcity, such as high population, climate change, carbonization, deforestation, mismanagement of water resources, natural calamities, and human greed. There is an urgent need to save water and find alternative methods to harvest water. In this regard, many researchers worked on bio-inspired designs based on beetles' water harvesting and transportation systems. In this chapter, many technological aspects of the bioinspired water harvesting mechanisms are discussed. The fabrication of such a graded surface for water collection and transportation is provided. Detailed modeling and simulation of wettability, transportation, and droplet growth are also discussed with a mathematical model. Several applications, representing bio-sensing, lab-on-chip, biomechanics, and energy conservation in solar systems, are discussed, along with the available advanced technology usage for further development to increase scalability to large production for efficiency and cost-effectiveness. The design should incorporate all the modules, such as droplet capture, condensation, and transportation. Durability and sustainability of these technologies must be given due attention before marketing these products. Stimuli-sensing, self-healing, and meta-materials hold great promise in these emerging technologies for water harvesting.

4.12 Future Research Challenges and Opportunities

Beetles' inspired water harvesting system is studied and researched well, but still, there are opportunities for improvement. The main challenge is to improve the durability of sensors, lab-on-chips, and other devices. These devices must operate consistently over the long term, giving good performance. Protection of textured surfaces from mechanical wear, thermal wear, chemical wear, and also UV radiation is an important design consideration. Temperature fluctuation is important in designing anti-icing systems. Sometimes, it is custom-made for a particular set of environmental parameters. Synthetic materials, meta-materials, and self-healing materials are promising material sets.

4.12.1 Adaptive Surface Technologies

Researchers are working on a stimulus-responsive polymer that can switch from hydrophobic to hydrophilic surfaces based on stimulus. This requires an extensive study on eco-friendly materials that can adapt to different weather conditions yet provide good feedback for correction and action. Smart materials and shape memory alloys can be designed as stimuli materials, which can change their properties with respect to temperature, fog density, and water, for the highest efficiency both in active and passive modes of operation.

4.12.2 Advancements in Manufacturing

Research efforts are now focused on many aspects of scalability for large lot manufacturing and on reducing the overall cost. Laser-based methods are expensive due to the high cost of the equipment, and alternative techniques for producing textures are to be explored, other than electrospinning. Recently, there has been tremendous growth in automation technology due to advances in artificial intelligence, machine learning, and deep learning. Expanding actions are the order of the day to produce green manufacturing, water conservation, and decarbonization. Lab-on-chip developments for cancer and other diseases can be done using these kinds of bio-inspired designs. Environmental protection is important, oil spills in the sea are common, and a water-oil separation mechanism based on bio-inspired design cannot only save marine life but also human beings. A multidisciplinary approach and networking with biologists, engineers, chemists, environmentalists, water conservation specialists, government agencies, and industries are important for the successful implementation of bio-inspired water harvesting systems.

References

1. Huang, Z.-X., Liu, X., Wu, J., Wong, S.-C., Qu, J.-P., Electrospinning water harvesters inspired by spider silk and beetle. *Mater. Lett.*, 211, 28–31, Elsevier, 2018.
2. Zhu, H., Cai, S., Zhou, J., *et al.*, Integration of water collection and purification on cactus- and beetle-inspired eco-friendly superwettable materials. *Water Res.*, 206, Article 117759, Elsevier, 2021.
3. Ju, J., Bai, H., Zheng, Y.-M., Zhao, T.-Y., Jiang, L., A multi-structural and multi-functional integrated fog collection system inspired by cactus. *Nat. Commun.*, 3, Article 1247, Springer Nature, 2012.
4. Zhang, S.-N., Huang, J.-Y., Chen, Z., Lai, Y.-K., Bioinspired special wettability surfaces: From fundamental research to water harvesting applications. *Small*, 13, 3, Article 1602992, Wiley-VCH, 2017
5. Wang, Y.-C., Zhang, L.-B., Wu, J.-B., Wang, P., A facile strategy for the fabrication of a bioinspired hydrophilic–superhydrophobic patterned surface for highly efficient fog-harvesting. *J. Mater. Chem. A*, 3, 37, 18963–18969, RSC, 2015.
6. Zheng, Y.-M., Bai, H., Huang, Z.-B., Jiang, L., Directional water collection on wetted spider silk. *Nature*, 463, 7281, 640–643, Springer Nature, 2010
7. Parker, A.R. and Lawrence, C.R., Water capture by a desert beetle. *Nature*, 414, 6859, 33–34, Springer Nature, 2001.
8. Chen, H.-W., Zhang, P.-F., Liu, H.-L., Jiang, L., Continuous directional water transport on the peristome surface of *Nepenthes alata*. *Nature*, 532, 7599, 85–89, Springer Nature, 2016.
9. Li, R.-Y., Shi, Y.-S., Wang, P., Harvesting water from air using anhydrous salt with sunlight. *Environ. Sci. Technol.*, 52, 9, 5398–5406, ACS, 2018.
10. Zhu, H., Yang, F.-C., Guo, Z.-G., High-efficiency water collection on biomimetic material with superwettable patterns. *Chem. Commun.*, 52, 84, 12415–12417, RSC, 2016.
11. Bai, H., Wang, L., Jiang, L., Efficient water collection on integrative bioinspired surfaces with star-shaped wettability patterns. *Adv. Mater.*, 26, 29, 5025–5030, Wiley-VCH, 2014.

12. Wang, Z., Lin, S., Zhu, Y., Jin, J., Nanoparticle-templated nanofiltration membranes for ultrahigh performance desalination. *Nat. Commun.*, 9, 1, Article 2004, Springer Nature, 2018.
13. Ju, J., Xiao, K., Bai, H., Jiang, L., Bioinspired conical copper wire with gradient wettability for continuous and efficient fog collection. *Adv. Mater.*, 25, 41, 5937–5942, Wiley-VCH, 2013.
14. Dorrer, C. and Rühe, J., Mimicking the *Stenocara* beetle: Dewetting of drops from a patterned superhydrophobic surface. *Langmuir*, 24, 12, 6154–6158, ACS, 2008
15. Zhu, H., Duan, R.-L., Wang, J.-H., Prewetting dichloromethane-induced aqueous solution adhered on Cassie superhydrophobic substrates to fabricate efficient fog-harvesting materials. *Nanoscale*, 10, 27, 13045–13054, RSC, 2018.
16. Jiang, L., Guo, C., Fu, M., Gong, X., Ramakrishna, S., Water harvesting on biomimetic material inspired by beetles. *Heliyon*, 9, Article e12355, Elsevier, 2023.
17. Choi, Y., Baek, K., So, H., 3D-printing-assisted fabrication of hierarchically structured biomimetic surfaces with dual-wettability for water harvesting. *Sci. Rep.*, 13, Article 10691, Springer Nature, 2023.
18. Wang, X., Zeng, J., Yu, X., Liang, C., Zhang, Y., Beetle-like droplet-jumping superamphiphobic coatings for enhancing fog collection of sheet arrays. *RSC Adv.*, 10, 3, 282–288, RSC, 2020.
19. Xu, C., Feng, R., Song, F., Wang, X., Wang, Y., Desert beetle-inspired superhydrophobic/superhydrophobic patterned cellulose film with efficient water collection and antibacterial performance. *ACS Sustain. Chem. Eng.*, 6, 11, 14679–14684, ACS, 2018.
20. Hu, R., Wang, N., Hou, L., Cui, Z., Liu, J., Li, Q., Zhang, H., Zhao, Y., A bioinspired hybrid membrane with wettability and topology anisotropy for highly efficient fog collection. *J. Mater. Chem. A*, 7, 10, 124–132, RSC, 2019
21. Liu, M., Peng, Z., Yao, Y., Yang, Y., Chen, S., Flexible functional surface for efficient water collection. *ACS Appl. Mater. Interfaces*, 12, 6, 12256–12263, ACS, 2020.

22. Li, X., Yang, Y., Chu, M., Sun, H., Shan, W., 3D-printed cactus-inspired spine structures for highly efficient water collection. *Adv. Mater. Interfaces*, 7, Article 1901752, Wiley-VCH, 2019.
23. Dolan, F., Lamontagne, J., Link, R., *et al.*, Evaluating the economic impact of water scarcity in a changing world, in: *Nat. Commun*, pp. 1–9, Springer, Dordrecht, 2021.

5

Biomimetic Bone Implant Design: Materials, Manufacturing, and Modeling

Abstract
Biomimetic bone implants will represent a state-of-the-art solution to the problems facing the bone repair and regeneration field, using advanced materials and fabrication techniques to mimic the structure and properties of natural bone effectively. A new paper assesses the feasibility of materials as titanium alloys, hydroxyapatite, and biocompatible polymers, considering the compatibility of mechanical, biological, and structural properties. Techniques in fabrication, such as additive manufacturing, rapid prototyping, and laser-assisted processes, are elaborately reviewed to highlight the merits of producing accurate and patient-specific implants. Further emphasis is placed on the role of finite element modeling in enhancing the stress analysis of implants during the design stage to optimize load-bearing and durability. The study incorporates material science, along with advanced manufacturing capabilities and computational simulations, thereby providing a much-needed overview of the biomimetic bone implants that currently exist and their future. These results are important to further activities in designing implants, which must balance both biological integration and mechanical functionality, a key thrust area to advance the field of regenerative medicine.

***Keywords*:** Biomimetic implants, material feasibility, fabrication techniques, finite element modeling, bone tissue engineering

Jagadeesha T. and Sandip Kunar. Bio-Inspired Manufacturing Systems, (109–134) © 2026 Scrivener Publishing LLC

5.1 Introduction

Generally, trauma and congenital abnormalities have remained major, consistent challenges in most global healthcare systems for bone injuries and defects, besides degenerative diseases. The skeleton gives a human being the body framework for bodily protection, structure, as well as a living structure in the execution of important functions like hematopoiesis. Hence, massive defects in the bones have remained very hard to correct because more than one demand exists about material properties and procedures [1]. Some traditional types are metal-based implants, titanium alloys, as well as stainless steel. They have come to find themselves crucial in offering support to structures. However, with these, several other negatives seem to stand in front of such implants.

Other problems are stress shielding, wherein the implants absorb too much load; hence, the adjacent bone does not experience enough mechanical stimulation, leading to bone resorption and loosening of the implants. Furthermore, the most commonly used materials for this application do not have an osteoinductive property, which is required to enable the implant itself to induce natural bone growth [2]. These issues demonstrate the need for new concepts that allow the integration of mechanical functionality with biological compatibility.

5.2 Biomimetic Bone Implants

The new direction towards the biomimetic bone implant, which is very similar in structure, properties, and functions to the natural bone, can be said to be a move ahead over the limits of traditional methods. Beginning with potential better integration into the body, fewer complications, and a better chance for tissue regeneration, biomimetic bone implantability arises from the ability to mimic the hierarchical organization and biochemical environment.

5.2.1 Materials for Biomimetic Bone Implant

The selection of the type of bone implants is the starting point for their success. Materials, which are selected for their use, should meet many conditions; for example, they should be biocompatible to ensure an excluded adverse reaction from a healthy body and have the closest possible similarity in all characteristics with natural bone [3]. Moreover, the material should stimulate bioactivity and cell adhesion for successful bone regeneration.

5.2.1.1 Classical Materials and Their Deficiencies

The most widely used is titanium and its alloys, which have good strength, corrosion resistance, and moderate biocompatibility. They do not have osteoinductive and bioactive properties desired for perfect integration.

5.2.1.2 New Developments and Hybrid Materials

The newest trends include ceramics, such as hydroxyapatite, bioglass, and biopolymers like polycaprolactone and polylactic acid. Their superior biocompatibility and bioactivity make it easier to interact with bone cells. Hybrid ceramics/polymer composites are a newly developed material. What these incorporate are the positive mechanical features of ceramics while still maintaining the flexibility of polymers, and above all, their degradability. Thus, advanced material science has introduced a new horizon in restoring structural integrity through restoration to co-function in healing [4].

5.3 Fabrication Methods

Manufacturing techniques have made giant leaps, whereby it is possible to have intricate and individually customized implants in technological advancements.

5.3.1 Additive Manufacturing

Additive manufacturing (AM) or 3D printing has taken the game by storm, providing precise control of scaffold geometry and porosity. This is essential for encouraging vascularization, cellular ingrowth, and nutrient exchange within the implant.

5.3.2 Other Techniques

Other techniques, such as electrospinning, can be used to produce nanofibers with properties that are equivalent to those of the electrochemical machining (ECM). They have significantly enhanced cellular adhesion and proliferation. Techniques range from laser-assisted sintering to rapid prototyping to prepare implants with tailored mechanical and biological properties.

5.4 Finite Element Modeling and Stress Analysis

Having determined the material and its fabrication, the mechanical integrity of an implant bears the following concerns:

5.4.1 Finite Element Modeling in Implant Design

Finite element modeling, or FEM, is a powerful computational tool applied for simulating the mechanical behavior of implants subjected to physiological loads. It can be understood by studying the stress distribution and deformation patterns; the model provides possible points of failure and directs the optimization of the design of implants [5].

5.4.2 Practical Applications and Benefits

For example, implants should distribute the applied loads evenly to avoid a concentration of stress that leads to implant failure or undesirable effects on the surrounding bone. Simulations using FEM allow scientists to test different geometries, materials, and conditions, hence developing designs that prove durable and compatible [6]. Furthermore, such analyses reduce the need to do long-term *in vivo* testing; therefore, hastening the development time without sacrificing safety and effectiveness.

5.5 Materials Design for Bone Implants

5.5.1 Criteria for Material Selection

The success of biomimetic bone implants largely depends on selecting suitable materials. These materials must satisfy several essential criteria to ensure both biological compatibility and mechanical functionality [7]. Each material must fulfil requirements related to its strength, biological response, and longevity. The following factors are critical when selecting materials for bone implants:

5.5.1.1 Mechanical Strength and Load-Bearing Capacity

The mechanical forces, especially in weight-bearing parts such as the femur, tibia, and spine, are substantial loads on bone implants. It is desired that materials have enough strength and resistance to deformation and mechanical stiffness in long-time exposure for them to serve as effective bone implants without failure or deformation. In particular, bone is a dynamic material with location-dependent mechanical properties; implants should be able to duplicate the load distribution noticed in natural bone. Moreover, under repetitive loading conditions, fatigue failure resistance is also required; otherwise, the bone implant could be subjected to chronic mechanical stress [8]. Titanium-based alloys are widely used in the aforementioned applications because of their high strength and resistance to fatigue failure. However, the mechanical properties of the implant material should correspond to those of the bone to prevent complications such as stress shielding, where the surrounding bone weakens from a lack of stimulation from the implant.

5.5.1.2 Biocompatibility and Osteointegration

Biocompatibility is the most important factor in implant selection, as it guarantees that the material does not trigger adverse immune reactions. Once the implant is introduced into the body, the surrounding tissue needs to accept and coalesce with it. Osteointegration is the method in which the bone directly couples to the implant, resulting in no additional surgical intervention for retention or structural support [9]. In this case, biomimetic implants should not produce any kind of inflammatory response to rejection by the immune system. Materials, such as titanium and hydroxyapatite, were found to be highly biocompatible, which supports the phenomenon of osteointegration without potential long-term complications from rejection or infection.

5.5.1.3 Bioactivity and Osteoinductivity

Besides being biocompatible, the material for a biomimetic bone implant must be bioactive as it plays an active role in the healing process by supporting adhesion, growth, and differentiation of cells. Osteoinductive properties are also very desirable; these materials stimulate stem cells to become bone-forming cells, therefore improving bone regeneration. Hydroxyapatite is one of the most common bioactive materials used for implants, especially in bone. This is because it can provide osteoconductivity, which is the property that allows the growth of bone along its surface [10]. Bioglass is another example of a bioactive material that exhibits strong chemical bonding with the bone, thus improving the process of osteointegration. However, despite these bioactive properties, many bioactive materials also have mechanical limitations; for example, brittleness that may make them unsuitable for high-stress load-bearing applications. The summary is given in Table 5.1.

Table 5.1 Criteria for material selection in biomimetic bone implants.

Criteria	Details	Example	Remarks
Mechanical Strength and load-carrying capacity	The femur, tibia, and spine must be able to carry a load	Titanium alloys	Stress shielding must be prevented
Fatigue resistance	Must have high fatigue resistance	Titanium and Co-Cr alloys	Good for spine implants
Biocompatibility	Neighboring tissues must accept	Titanium and Hydroxyapatite	Important for patient safety
Osteointegration	Direct bonding	Porous coated materials	Surface treatment and porosity are parameters
Bioactivity	Should support adhesion	Hydroxyapatite and bioglass	For fixing
Osteoconductive	The bone should grow	Hydroxyapatite/ Bioglass and coated ceramics	Regeneration and reconstruction of bones
Corrosion resistance	Must resist degradation	Titanium and tantalum	Prevents the release of toxic ions for durability
Brittleness	Must not be brittle. Some ceramics are brittle	Hydroxyapatite and bioglass	Non-load-bearing and coated application
Elastic Modulus	Must match bone stiffness	PEEK composites	Properties must be near the cortical bone
Wear resistance	Should not produce debris	Co-Cr alloys and UHMWPE	Debris can lead to osteolysis

(*Continued*)

Table 5.1 Criteria for material selection in biomimetic bone implants. (*Continued*)

Criteria	Details	Example	Remarks
Surface modification	Must be bioactive and coatable	Titanium anodized or composite coating	Improves cell response
Manufacturability	Must be easily fabricated	Ti-6Al-4V PEEK	Additive manufacturing
Cost effectiveness	Should be economical and affordable	Stainless steel	Performance and affordability must be considered

5.5.2 Degradability and Controlled Resorption

For temporary bone implants, materials that degrade slowly over time are often desirable. These degradable materials are eventually resorbed by the body, eliminating secondary surgeries to remove the implant. This means that the degradation rate of the material should be aligned with the bone formation rate. If it degrades too quickly, this will cause implant failure; if the degradation is too slow, a cavity forms within the bone, leading to delayed healing and instability. Biodegradable polymers, including polycaprolactone (PCL) and polylactic acid (PLA), have been among the various materials studied and utilized as bone implants [11]. They have a controlled degradation rate, allowing these materials to be used in such implants where replacement by natural bone is achieved upon progressive healing.

5.6 Materials Combinations for Implant

Several materials are commonly used for bone implants due to their favorable mechanical properties, biocompatibility, and osteointegration abilities. These materials have been extensively studied and are widely used in both clinical and experimental settings. The most commonly used materials include titanium, hydroxyapatite, bioglass, and polymers [12].

5.6.1 Titanium and Titanium Alloys

Titanium alloys, especially Ti-6Al-4V, are the most commonly used materials for bone implants because they offer a combination of high strength, light weight, corrosion resistance, and biocompatibility. These materials have been used in orthopedic implants, dental implants, and spinal fusion devices. However, titanium alloys with superior mechanical properties have poor osteointegration potential. Titanium alloys are often coated with bioactive materials, such as hydroxyapatite, for enhancing their bioactivity, thus improving bonding between the implant and surrounding bone [13]. Titanium also has very high corrosion resistance, which allows it to be used as an implantable material for a long period. Still, since titanium does not activate bone growth, surface treatments such as plasma spraying, electrochemical coating, or acid etching are usually applied to improve its bioactive properties. Figure 5.1 shows the use of titanium and titanium alloys in dental implants.

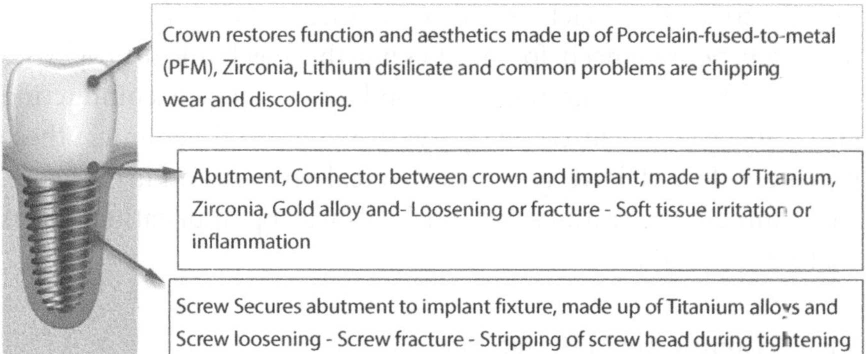

Figure 5.1 Dental implant parts with material type and common problems.

5.6.2 Hydroxyapatite and Bioglass

Hydroxyapatite (HA) is a naturally occurring mineral that makes up the inorganic part of bone; it is very bioactive, supporting bone formation by serving as a scaffold for attaching and proliferating bone cells. These HA coatings are applied to metal implants to enhance their osteointegration properties. However, HA is brittle and mechanically weak; hence, its application as a load-bearing implant component is limited and usually reinforced [12, 14]. Another bioactive material is bioglass, a silica-based material, which strongly bonds with the bone tissue and improves osteointegration. While both HA and bioglass are excellent for promoting bone growth, their brittleness limits them to applications that require a lot of mechanical strength.

5.6.3 Polymers (PCL and PLA)

Biodegradable polymers, such as PCL and PLA, are used in bone implants where temporary support is required. These materials are biocompatible and biodegradable. Thus, they can ideally be used in applications where the implant will degrade as the bone heals. Both PLA and PCL have controlled degradation rates that can be engineered to particular time frames to ensure that the implant remains intact until the bone has healed sufficiently. By incorporating bioactive fillers, such as hydroxyapatite or collagen, these polymers can be enhanced for mimicking the mechanical properties of bone and for promoting osteoconductivity and osteoinduction [15]. However, when compared with metals and ceramics, these polymers do not have good mechanical strength and, therefore, have limited applications in non-load-bearing applications or as coatings or scaffolds.

5.6.4 Composite Materials

Composites that have gained much more attention in bone implant design are those that incorporate the advantages of various material classes. The composites incorporate ceramics, such as hydroxyapatite, with biodegradable polymers, such as polycaprolactone, to produce implants. These implants possess an amalgam of mechanical strength, bioactivity, and degradability [16]. Such composites, although superior in their mechanical properties, still maintain the osteoconductive and bioactive properties of the ceramic components. For instance, HA/PCL composites provide an excellent microenvironment for bone cell growth since they have a strength that is large enough to tolerate the mechanical stress of the healing process. Such a combination of properties makes hybrid materials promising in the application area of load-bearing bone implants with high strength and biological activity correlated.

With advanced nanotechnology, nanoscale coatings, which improve the surface characteristics of bone implants, have been developed. Such coatings may be added onto the titanium or polymer-based implants, enhancing bioactivity and promoting osteointegration. The covalent attachment of extracellular matrix proteins, such as collagen, fibronectin, or other molecules, has proven to promote cell adhesion, growth, and differentiation on the implant surface. In addition, they can be designed and engineered for the delivery of either controlled or otherwise growth factors or drugs, which subsequently promote the healing process. The most favored in depositing these films are the plasma spray process, chemical vapor deposition method, and the sol-gel process since they appreciably enhance the bioactivity of the implant material from which they are derived [17]. Such nanocoating may be designed with interesting tunable properties at the nanoscale level and may lead to promising opportunities in implant designing for better interaction with biological tissues.

5.6.5 Smart Materials and Functionalized Scaffolds

Smart materials are a relatively recently known class of materials sensitive to environmental stimuli, such as pH, temperature, or mechanical stress. These materials can produce functionalized scaffolds that release bioactive agents, growth factors, or cytokines, according to the change in the local environment. For example, some smart scaffolds can deliver osteoinductive growth factors to those regions of the implant site that experience mechanical stress. This helps the site to have bone formation. Scaffolds can be engineered to degrade at different rates and dynamically respond to the healing process. The future of biomimetic bone implants is found in designing smart materials capable of active change under the conditions to meet their environment [18]. Additive manufacturing or 3D printing changed all the rules about the production of bone implants altogether: it is now possible to manufacture customized complex geometries up to near-microarchitectural resolutions, thus making bone implants close to their native microarchitecture. It also gives highly important control over the porosity, surface area, and material distribution of the implant, so that it allows cellular infiltration and nutrient exchange. Hybrid or bioactive materials in 3D printing allow implants to fit the specific needs of individual patients, while also achieving structural and biological compatibility. The HA/PCL scaffolds can be 3D printed with controlled porosity that promotes bone ingrowth and vascularization. The approach of personalized design for

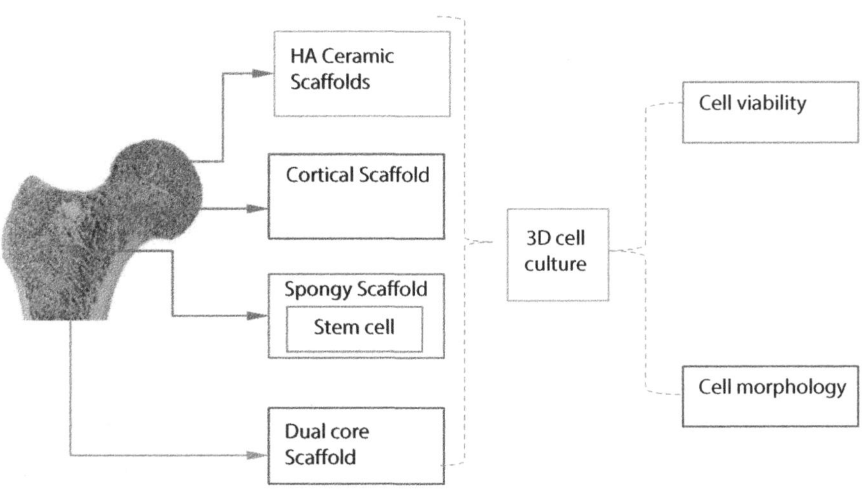

Figure 5.2 Various scaffolds in the bone.

implants holds immense potential for improving patient outcomes and reducing the rate of complications. Various scaffolds are shown in Figure 5.2.

5.7 Fabrication Techniques

The fabrication of biomimetic bone implants has seen tremendous progress, shifting from traditional techniques to modern technologies. Such techniques are of paramount importance in replicating the complex structure, mechanical properties, and bioactivity of natural bone to address the challenges of bone repair and regeneration. This session discusses various fabrication methods, their principles, applications, and associated challenges.

5.7.1 Additive Manufacturing

Additive manufacturing is broadly termed as 3D printing. It has changed the scope of implant fabrication with unmatched customization and precision. This process can, layer by layer, construct geometries with complex details in mimicking both macroscopic and microscopic bone features.

5.7.1.1 Key Attributes of Additive Manufacturing

1. Accuracy and flexibility: AM allows the addition of micro- and nano-architectures that should be added to replicate the hierarchical nature of bone. Such accuracy will allow higher cellular incorporation along with vascularization due to pore size, shape, and interconnectivity.

2. Biomaterial choice: PCL, PLA, HA, and titanium alloys are some of the most frequently utilized biomaterials, though hybrids combining polymers and ceramics have become increasingly popular for the optimization of bioactivity and mechanical strength.

3. Personalization: Patient-specific implants can be designed to keep the medical imaging modalities, say CT or MRI data of the patient, such that the implants can be directly fabricated from the defect location in case of craniofacial defect cases.

5.7.1.2 Applications

Scaffold Design: Application-specific design tasks performed by 3D printed scaffolds, for instance, in the case of maxillofacial reconstruction and spinal implant scaffold, porous networks providing cellular nutrient exchange.

Bone Regeneration Research: AM scaffolds are used in preclinical studies to study new biomaterials for their potential in osteoconductive and osteoinductive properties.

5.7.1.3 Limitations

Additive manufacturing has poor availability of bioactive materials for printing, poor mechanical strength for load-bearing applications, and time-consuming, complex builds, though it has many advantages.

5.7.2 Traditional Techniques

Traditionally, casting and sintering alongside mold-based techniques have been employed within the production of implants, wherein most of them have been generic for decades. These do not embrace many specific and flexible AM methods, though significant for the chosen material in conjunction with respective application [19]. Cast and sintered, these methods often arise when producing metal and sometimes ceramic parts. The titanium and cobalt-chrome alloys used are generally chosen for strength and anticorrosive capability. Forming: Generally, all polymer-based prosthetics are formed through either injection or compression molding. This is the economical and scalable means of doing it.

5.8 Stress Analysis and Finite Element Modeling Simulation

This is because mechanical reliability is an essential attribute of implants, so that biomimetic bone implants will operate when required in load-carrying applications. Therefore, an important tool for this purpose would be a stress analysis coupled with FEM to gain insight and optimize the mechanical behavior of implants. The mechanical behavior of these implants can be predicted before how much they would compose with bone and also oppose physiological loads toward creating an insight handy in choosing the material, fine-tuning designs, and clinical applications [20].

This chapter will cover relevance to stress analysis, methodology involved in FEM simulations, significant case studies, and areas left for hands-on contribution.

5.8.1 Stress Analysis Role in Implants

The mechanical demands on bone implants vary considerably, both by location and purpose. For instance, femoral or spinal implants must withstand very high compressive forces, torsional stress, and fatigue cycles. Implants located in the cranial area have a whole different mechanical environment [21]. With the use of any implant, an element of stress analysis is necessitated so that implants:

1. Reproduce bone load-bearing capability: it should approximate bone without creating conditions known as stress shielding, whereby bones take up excess load borne by the implant and resorb it.
2. Prevention of failure: the high stress concentration, which may lead to material fatigue or micro fracture, and prevention of the same through design.
3. Long-term compatibility: Implants must endure cyclic loading over a long period of time with no failure or degradation.

For example, titanium-based implants have been one of the best for their use in load-bearing applications concerning the strength-to-weight ratio, but unless optimally stressed, they are still prone to introducing stress shielding and other problems.

5.9 Mathematical Modeling and Analysis

Mathematical models serve as the most important step in understanding the performance of biomimetic bone implants. Models predict stress distribution and optimize scaffold designs, thus considering complex biological interactions. Equations used for scaffold mechanics, thermal modeling, porosity control, and finite element analysis provide detailed insights, helping researchers improve implant performance [22]. Each mathematical approach, supported by illustrative examples and detailed derivations, is based on published research that provides a solid foundation for improving the integration and functionality of bone implant designs.

5.9.1 Mechanical Behavior and Stress Analysis

5.9.1.1 Fundamental Stress-Strain Relationships

The performance of bone implants depends on their ability to withstand mechanical loads without failure. The basic relationship between stress (σ) and strain (ε) is given by:

$$\sigma = E \cdot \varepsilon$$

where:

σ = Stress (MPa),
E = Young's modulus (MPa),
ε = Strain (dimensionless).

For materials with elastic and plastic deformation, the total strain is expressed as:

$$\varepsilon = \varepsilon_e + \varepsilon_p$$

where:

ε_e = Elastic strain,
ε_p = Plastic strain.

5.9.1.2 *Bending Stress in Cylindrical Implants*

When a cylindrical implant experiences bending, the maximum stress is calculated as:

$$\sigma_b = \frac{M \cdot c}{I}$$

where:

σ_b = Bending stress (MPa),
M = Bending moment (N·mm),
c = Distance from the neutral axis (mm),
I = Moment of inertia (mm^4).

Example:
For an implant with $M = 500\ N \cdot mm,\ c = 2\ mm,$ and $I = 20\ mm^4$:

$$\sigma_b = \frac{500 \cdot 2}{20} = 50\ MPa$$

5.9.1.3 *Stress Distribution in Porous Scaffolds*

Porous scaffolds must balance mechanical strength with biological functionality. The stress distribution (σ) for stochastic scaffolds is influenced by their pore geometry and can be modeled using:

$$\sigma = \sigma_0 (1 - P)^n$$

where:

σ_0 = Stress of the solid material,
P = Porosity fraction,
n = Material-dependent constant.

For trabecular bone mimetics with $P = 0.6$, $n = 2.2$, and $\sigma_0 = 200$ MPa:

$$\sigma = 200(1-0.6)^{2.2} = 35.84\ MPa$$

5.9.1.4 *Yield Strength in Implants*

The yield strength (σ_y) is critical to ensure the implant withstands physiological loads without permanent deformation:

$$\sigma_y = \frac{\sigma}{SF}$$

where:

SF = Safety factor.

For implants, SF typically ranges between 1.5 and 2.0 to ensure reliability under varying conditions.

5.9.2 Thermal and Fabrication Modeling

Thermal and fabrication modeling play key roles in the application of bio-inspired design, particularly in the case of tissue engineering scaffolds. For instance, rapid prototyping-based fabrication techniques allow for strict control of scaffold architecture, leading to pore sizes and geometries to mimic biological tissues [23]. Further, thermal processes, such as oxygen plasma modification, have been shown to improve surface properties of materials like PCL for better attachment and cell proliferation. These techniques have a balance between material properties and biological compatibility, ensuring that scaffolds not only replicate natural designs but also function effectively in medical applications [24].

5.9.2.1 *Heat Conduction in Scaffold Materials*

During additive manufacturing, heat transfer is modeled using the Fourier heat conduction equation:

$$\frac{\partial T}{\partial t} = \alpha \nabla^2 T + \frac{q}{\rho c}$$

where:

$\alpha = \frac{k}{\rho c}$,
q = Internal heat generation per unit volume (W / m^3),
k = Thermal conductivity $(W / m \cdot K)$,
ρ = Density (kg / m^3),
c = Specific heat$(J / kg \cdot K)$.

5.9.2.2 *Thermal Stress in Implants*

Thermal stresses arise due to uneven temperature distributions:

$$\sigma_T = \alpha_T E \Delta T$$

where:

α_T = Coefficient of thermal expansion,
ΔT = Temperature change.

Example for titanium $(E = 110\ GPa,\ \alpha_T = 8.6 \times 10^{-6}\ K^{-1})$ with $\Delta T = 50\ K$:

$$\sigma_T = 8.6 \times 10^{-6} \cdot 110 \cdot 50 = 47.3\ MPa$$

5.9.3 Biological and Remodeling Models

Biological and remodeling models are important in the design of biomimetic bone implants. They should mimic the regenerative processes and structural adaptation of natural bone. The papers indicate that scaffolds mimicking the extracellular matrix use biomaterials, such as collagen, hydroxyapatite, and β-tricalcium phosphate (β-TCP), to support osteoconduction and osteoinduction. Remodeling models focus on the dynamic interaction between cells and materials, and osteoblast and MSC activities drive bone deposition and mineralization [25]. For instance, protein-coated scaffolds improve cell adhesion and differentiation to facilitate uniform extracellular matrix formation. Moreover, advanced designs integrate hierarchical structures that guide tissue regeneration while gradually degrading at rates equal to natural bone turnover. These models ensure not only that the biomimetic implants replace the damaged bone but also stimulate their continuous remodeling, thus achieving long-term functional integration [26].

5.9.3.1 Bone Adaptation Under Stress

Bone remodeling follows Wolff's Law, where density adapts to stress:

$$\Delta\rho = k(\sigma - \sigma_t)$$

where:

$\Delta\rho$ = Change in bone density,
k = Remodeling rate constant,
σ = Applied stress,
σ_t = Threshold stress.

5.9.3.2 Cellular Growth in Scaffolds

Cell proliferation in scaffolds is modeled using logistic growth:

$$\frac{dN}{dt} = rN\left(1 - \frac{N}{K}\right)$$

where:

N = Cell number,
r = Growth rate,
K = Carrying capacity.

5.10 Porosity Design and Optimization

Porosity design and optimization are important factors in scaffold-based tissue engineering, with implications for cell distribution, proliferation, and tissue formation. Studies suggest that 3D bio-plotted scaffolds, as those fabricated with polycaprolactone, exhibit interconnected pore architectures and porosity tunable within a range of 60% to 78%. This design ensures uniform cell colonization and nutrient diffusion throughout the structure [27]. In contrast, conventional scaffolds with irregular, compact architectures often restrict tissue growth to peripheral regions [28]. Optimizing pore size, for example, to 300 μm in bio-plotted scaffolds, balances mechanical stability and biological compatibility for effective mimicry of native tissue environments. Such controlled porosity promotes uniform extracellular matrix deposition and supports successful tissue regeneration.

5.10.1 Level-Set Equations for Porosity Control

Porosity gradients are designed using level-set functions:

$$\phi(x,y,z) = \cos\left(\frac{2\pi x}{L}\right) + \cos\left(\frac{2\pi y}{L}\right) + \cos\left(\frac{2\pi z}{L}\right) - d$$

where:

L = Unit cell size,
d = Porosity control parameter.

5.11 Conclusion

The articles under consideration collectively emphasize the importance of scaffold design, the materials used in fabrication, and the biological response in the field of bone tissue engineering. Scaffolds are absolute frameworks; their architecture determines cell attachment, growth, and overall distribution and uniformity of the developed tissue. However, while open-pore structures such as plotted PCL scaffolds enable high levels of colonization, their differentiation induction is slower than that of compact scaffolds. Such strategies that are employed include the use of protein-based coatings, such as collagen and fibronectin. When added to the osteogenic differentiation culture, tissue compositional homogeneity increases. However, most of the time, such improvements are accompanied by some disadvantages concerning early-stage differentiation. The incorporation of bioactive materials such as hydroxyapatite and beta-tricalcium phosphate also serves to make the material more bone-like, encouraging osteoinduction and osteoconduction. The adipose-derived stem cells (ADSCs) and stem cells are also useful for enhancing regeneration when coated onto non-grafts, greatly assisting with the even formation of tissues, as well as helping with robust differentiation of osteogenic lineages. However, mechanical compatibility is the most important as it will limit stress shielding and allow good integration of the scaffold. Novel approaches regarding hybrid scaffolds for the development of bone or bone-like structures consisting of a desired shape with bioactive surface coating are considered to be very useful. These studies underscore the need for a balanced approach that harmonizes scaffold architecture, bioactivity, and cellular dynamics to overcome current limitations in bone tissue engineering.

5.12 Scope for Future Work

Future work can target the design of scaffolds to combine uniform tissue distribution with favorable differentiation timing using hybrid architectures. Advanced surface modifications will include nanoscale topographies or growth factor integration and can enhance cell attachment and proliferation [29]. A dynamic culture system, such as a bioreactor, can closely mimic physiological conditions to achieve rapid cellular colonization and maturation of the extracellular matrix. Further, BMP and VEGF incorporation can help in the healing process through bioactive molecules, BMPs, and VEGF. Long-term *in vivo* studies will be helpful in ascertaining scaffold integration and functionality into the bone defects. Functional scaffolds with osteogenic, anti-inflammatory, and antimicrobial properties will further reduce complications. Additional engineering of ADSCs to possess greater osteogenic capacity can further alleviate limitations within the paradigm of tissue engineering [30]. These achievements may revolutionize the efficacy of scaffold-based bone regeneration for clinical applicability.

References

1. Lin, Y.C. and Lee, H.S., Optimization of machining parameters using magnetic-force-assisted EDM based on gray relational analysis. *Int. J. Adv. Manuf. Tech.*, 42, 1052, 2009.
2. Lin, Y., Chen, Y.F., Wang, D.A., *et al.*, Optimization of machining parameters in magnetic force assisted EDM based on Taguchi method. *J. Mater. Process. Tech.*, 209, 3374, 2009.
3. Ponomarev, A.V., Yakukhin, V.G., Maksimov, A.D., Hybrid machining technologies. *Russ. Eng. Res.*, 31, 717, 2011.
4. Zhang, Y.F., Li, L.W., Chen, J.M., Biomimetic design of bone implants with functionally graded porous structures. *J. Biomech. Eng.*, 141, 081001, 2019.
5. Bose, S., Roy, M., Bandyopadhyay, A., Recent advances in bone tissue engineering scaffolds. *Trends Biotechnol.*, 30, 546–554, 2012.
6. Sachlos, E. and Czernuszka, J.T., Making tissue engineering scaffolds work. Review on the application of solid freeform fabrication technology to the production of tissue engineering scaffolds. *Eur. Cell Mater.*, 5, 29–39, 2003.

7. Ngo, T.D., Kashani, A., Imbalzano, G., *et al.*, Additive manufacturing (3D printing): A review of materials, methods, applications and challenges. *Compos. Part B Eng.*, 143, 172–196, 2018.
8. Kurtz, S.M. and Devine, J.N., PEEK biomaterials in trauma, orthopedic, and spinal implants. *Biomaterials*, 28, 4845–4869, 2007.
9. Li, P. and Liu, X.Y., Hydroxyapatite and its application in bone repair. *J. Mater. Sci. - Mater. Med.*, 24, 2311–2323, 2013.
10. Rai, S.K., Mishra, R.K., Basu, B., Bioinspired design of bone-mimetic implants: A mechanobiological perspective. *Acta Biomater.*, 125, 1–18, 2021.
11. Zadpoor, A.A., Mechanics of additively manufactured biomaterials. *J. Mech. Behav. Biomed. Mater.*, 70, 1–6, 2017.
12. Niinomi, M., Mechanical properties of biomedical titanium alloys. *Mater. Sci. Eng., A*, 243, 231–236, 1998.
13. Sawant, R.R. and Kessler, M.J., Finite element modeling of load distribution in orthopedic implants. *ASME J. Med. Devices*, 14, 031002, 2020.
14. Yang, F. and Baldwin, M.W., Topology optimization for patient-specific bone implants. *Struct. Multidiscip. Optim.*, 62, 345–358, 2020.
15. Melchels, A.J., Feijen, J., Grijpma, D.W., A review on stereolithography and its applications in biomedical engineering. *Biomaterials*, 31, 6121–6130, 2010.
16. Gaharwar, A.K., Singh, I., Khademhosseini, A., Engineered biomaterials for *in situ* tissue regeneration. *Nat. Rev. Mater.*, 5, 686–705, 2020.
17. O'Brien, C.M., Paredes, C.M., Bowles, R.D., *et al.*, Three-dimensional printing of bone repair scaffolds using a biomimetic approach. *Biofabrication*, 8, 015003, 2016.
18. Van der Stok, J., Li, Z., Verdonschot, N., *et al.*, Selective laser melting-produced porous titanium scaffolds regenerate bone in critical size cortical bone defects. *J. Orthop. Res.*, 31, 792–799, 2013.
19. Hutmacher, D.W., Scaffolds in tissue engineering bone and cartilage. *Biomaterials*, 21, 2529–2543, 2000.
20. Elias, C.N., Lima, M.I., Barreto, L.M.S., Biomedical applications of titanium and its alloys. *JOM*, 60, 46–49, 2008.
21. Deckers, H.A., Dansen, T., Dubroel, P., Cornelissen, M.J., The role of scaffold architecture and composition on the bone formation by adipose-derived stem cells. *Tissue Eng. Part A*, 20, 1–2, 2014.

22. Lee, H. and Kim, G.H., Three-dimensional plotted PCL-b-TCP scaffolds coated with a collagen layer: Preparation, physical properties and *in vitro* evaluation for bone tissue regeneration. *J. Mater. Chem.*, 21, 520–529, 2011.
23. Gómez-Barrena, E., Rosset, P., Lozano, D., *et al.*, Bone fracture healing: Cell therapy in delayed unions and nonunions. *Bone*, 72, 1–9, 2015.
24. Malizia, P., Davis, S., Su, B., From bone to nacre–development of biomimetic materials for bone implants: A review. *Biomater. Sci.*, 2, 443–461, 2014.
25. Kumari, S., Ali-Ketan, G., Vijayavenkataraman, S., A novel method to design biomimetic stochastic scaffolds with controlled porosity for bone tissue engineering. *Mater. Des.*, 220, 110857, 2022.
26. Oladepo, B., Adekaiye, A., Ismail, S., *et al.*, 3D-printed biomimetic bone implant polymeric composite scaffolds. *Int. J. Adv. Manuf. Technol.*, 126, 4295–4217, 2023.
27. Sladkova-Faure, M., Dupri-Palmer, M., Othon-Angui, C., *et al.*, A biomimetic engineered bone implant for high accuracy of prosthetic implants. *Sci. Rep.*, 10, 22154, 2020.
28. Muller, M.J., Hasan, F.F., Horton, F.T., Strapi, R., Bone-implant interface shear modulus and stiffness of cortical-assisted rabbit model of open-pore Ti6Al4V. *J. Appl. Mech.*, 39, 65–72, 2020.
29. Oladepo, B., Ikumapayi, O., Obasius, A.S., Olowole, B., Adebiyi, V.A., Usman, H., Khan, A., Mechanical characterization of a polymer scaffold for bone implant materials for life sciences. *J. Mater. Sci.*, 55, 9057–9069, 2020.
30. Oladepo, B.I., Zahedi, S.A., Chong, S., Omigbodun, F.T., Malachi, I.O., 3D printing of surface characterization and finite element analysis improvement of PEEK-HAP-GO in bone implant. *Int. J. Adv. Manuf. Technol.*, 106, 829–841, 2020.

6

Gecko-Inspired Adhesives: A Multiscale Approach to Biomimetic Adhesion and Applications

Abstract

Geckos walk upside down due to tiny chain-like setae on their feet that form hierarchical, multiscale structures under van der Waals forces for adhesion. These structures possess unique properties, such as self-cleaning, tunable releasability, and effectiveness on wet and rough surfaces. Research shows that gecko adhesion works through friction, where shear forces remove microstructures with minimal peeling, challenging traditional tape-like adhesive models. The anisotropic properties of gecko setae and their keratin-based architecture have inspired the development of synthetic adhesives with improved adhesion and durability. In this chapter, the design principles behind gecko-inspired adhesives and how their structural attributes have been translated into synthetic materials are discussed. This chapter also presents various applications of these adhesives in fields like robotics, medical devices, and wearable technologies. Robots using synthetic dry adhesives can climb vertical and inverted surfaces, providing solutions for inspection, repair, and exploration in inaccessible areas. Soft robotic grippers now incorporate bioinspired gecko adhesives, enhancing strength and motion range while reducing energy costs. Medical adhesives modeled after geckos improve tissue adhesion, biocompatibility, and biodegradability, advancing wound closure and drug delivery systems. Studies also highlight the evolutionary and ecological aspects of gecko adhesion. Phylogenetic evidence suggests that adhesive toe pads evolved independently multiple times, revealing structural diversity and evolutionary transitions in this complex system. Habitat interactions further influence adhesive performance, inspiring innovations in biomimetic technology.

***Keywords*:** Gecko-inspired adhesives, van der Waals forces, synthetic adhesives, biomimetic technology, evolutionary adhesion

Jagadeesha T. and Sandip Kunar. Bio-Inspired Manufacturing Systems, (135–162) © 2026 Scrivener Publishing LLC

6.1 Introduction

Geckos, a lizard species, have an extraordinary ability to climb vertical surfaces and upside down on ceilings due to the biological structure of their feet. This ability enables them to adhere to surfaces without any use of liquids or any adhesive particles [26]. This unique occurrence has inspired research in biomimicry, where engineers mimic this to solve many challenges that humans face. From these challenges, adhesives that mimic the properties of gecko feet were developed [19].

Gecko-inspired adhesives offer tremendous potential across a wide range of fields. This technology offers vast innovations for many purposes, from medical applications such as surgical tapes to robotic grippers capable of climbing walls. However, replicating this gecko adhesion in present materials is challenging [19]. After all, it looks ahead in the prospects, highlighting the transformative impacts that adhesion could have in medicine, science, and other industries [27].

6.2 Historical Background

One of the ancient Greek philosophers and naturalists, Aristotle, marveled at their climbing powers but lacked the tools to uncover the mechanism hidden from them. Many scientists began to look for the role of physical and biological aspects of their adhesive abilities. In the mid-20th century, modern scientific exploration of gecko adhesion began. They believed that geckos used suction and sticky secretions to adhere to surfaces, but after many findings, these theories were debunked [6]. In the 1990s, Kellar Autumn and his team used advanced microscopy behind gecko adhesion and revealed the true mechanism behind gecko adhesion. They discovered that setae, a microscopic hair-like structure, is the main reason for the gecko adhesion [28].

6.3 Scientific Relevance of Gecko Adhesion

Gecko adhesion, a phenomenon paramount to biology, physics, material sciences, and engineering, stands for the development of new challenging materials. These demonstrate that even weak molecular interactions have enormous potential to generate usable manifestations since they add up *via* the van der Waals interactions created by millions of microscopic setae and nanoscale features with which geckos attach themselves to a surface. This adhesion mechanism is fascinating as it is dry, reversible, and self-cleaning; this has driven applications in biomimetic materials to develop environmentally friendly and reusable synthetic adhesives in a variety of fields [6].

In the field of nanotechnology, gecko feet have become a development in microfabrication and nano-structuring techniques that led to designing materials that feature high toughness and flexibility. Robotics has also greatly benefited, like grippers and climbing robots inspired by geckos for innovations in search and rescue missions and space exploration [6].

6.4 Significance of Biomimetic Studies

Biomimicry, which emulates designs in nature to solve human problems, has become increasingly popular in the last few decades. Such studies have deepened our understanding of how to optimize adhesive performance while elaborating ecological and material efficiency [4, 17]. Gecko-inspired materials have gone beyond adhesion, inspiring designs for ownable technologies, soft robotics, and medical devices [20, 21]. In addition, biomimetic studies stimulate new fabrication methods such as nano-structuring and material synthesis, leading to scaled manufacturing of high-performance adhesives [12]. As the gecko uses its high surface area, articulating setules to leverage molecular forces, these studies address real-world problems while enabling advances in fundamental science.

6.5 Objective of the Review

The main objective of this review is to summarize the advances in gecko-inspired adhesives from a holistic perspective, covering structural, as well as functional aspects, and their interrelation. The review focuses on principles governing gecko adhesion, including contact splitting, surface energy modulation, and mechanical compliance [16]. It also surveys cutting-edge fabrication techniques that have converted biological insight into marketable products, such as microfabricated adhesives and polymeric nanostructures [19]. Lastly, this review covers some of the diverse applications of adhesives in the field of robotic grippers and medical sealants, showcasing the transformative ability of these materials on the landscape of science and technology today.

6.6 Structure-Function Relationship

This hierarchical organization of the gecko toe pad is an example of the interplay between bio-structure and bio-function. At the macro-level, gecko toes are outfitted with scansors — flattened structures that contain millions of setae. These structures are insoluble keratin [1, 3], with a compliant but stiff mechanical behavior, important for working against the rough surfaces they encounter. At the nanoscale, the spatulae maximize intermolecular interactions through intimate contact with the substrate, and use van der Waals forces to create a strong yet reversible adhesive effect [23]. The principle of splitting up the contact on larger contacts into smaller and independent units [17] improves adhesive reliability and minimizes work done by disbanding the contact.

6.7 Mechanisms Underpinning Gecko Adhesion

Gecko adhesion is remarkably efficient due to a well-tuned interplay of forces – van der Waals interactions, capillary forces, and mechanical interlocking, among others. In dry adhesion, van der Waals forces dominate; this allows geckos to stick to smooth surfaces in the absence of liquid intermediates [1, 23]. Capillary forces are also present, particularly in wet environments, where liquid bridges may assist adhesion [8, 9]. Furthermore, the compliance of gecko setae enables them to conform to the geometries of surfaces, and their unipolar alignment allows them to be detached relatively easily under shear forces [5]. Such mechanisms emphasize the evolutionary optimization of gecko adhesion across environmental continua. The micro- and nanostructures are present in the gecko's feet, as shown in Figure 6.1

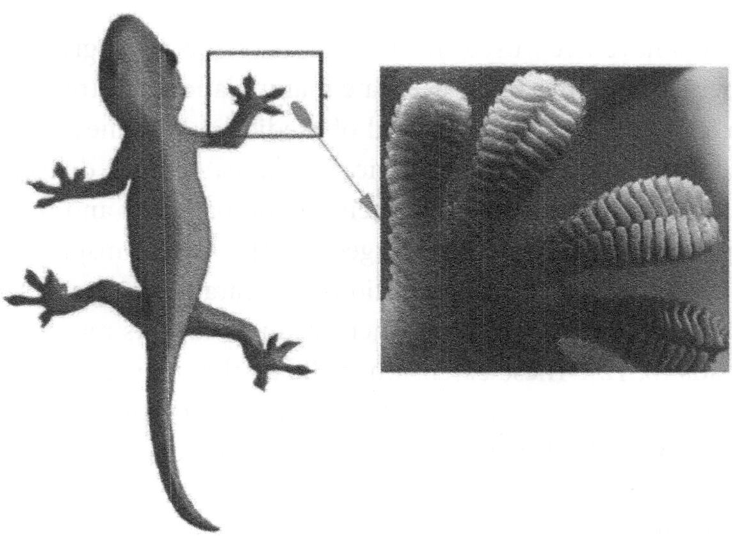

Figure 6.1 Micro- and nanostructures of gecko feet.

6.8 Fabrication Techniques for Gecko-Inspired Adhesives

To translate gecko adhesion to synthetic materials, sophisticated fabrication techniques have to be developed. Techniques of microfabrication and nano-structuring have been instrumental in mimicking the hierarchical structure of gecko setae. Nanoscale features can also be formed using photolithography, soft lithography, and electrospinning, generating fibrillar structures at the nanoscale [4, 12]. The selection of materials is also important, and common choices such as polydimethylsiloxane, polymers like silicone, and polyurethane that are elastic and biocompatible have been widely used [21, 22]. Recent advances have incorporated functional coatings that improve both wet adhesion and durability, allowing the design of multifunctional adhesive systems [8].

6.9 Potential Applications

Gecko-inspired adhesives are applicable in a wide range of fields, highlighting their versatile and game-changing nature. The same adhesives have been used in the context of robots in grippers and climbers, where they allow accurate manipulation and travel on difficult terrain [24, 25]. In the field of medicine, bioadhesives can be utilized in tissue adhesives, surgical sealants, or drug delivery systems, and it is noteworthy that their biocompatibility and reversible adhesion properties are advantageous [21]. Gecko-inspired adhesives are investigated for applications in wearable electronics, space exploration, and even as consumer products such as reusable tapes and pads [8, 12]. These extensions validate the interventional reach of gecko structures, merging literally and theoretically biological practices of applications with engineering practices of development. Table 6.1 summarizes the various applications and main benefits with examples.

Table 6.1 Summary of various applications.

Application area	Descriptions	Examples
Robotics	Used in grippers and climbers due to their great adhesive properties	Wall-climbing robot
Medical	Used as tissue adhesives and drug delivery patches	Wound closure patches
wearable electronics	Flexible and integration with skin contact devices	Health monitors
Space	Used for zero gravity conditions	Dry, adhesive gripping systems
Consumer product	Tapes, adhesives, pads	Wall-hanging pads and cable organizers
Micromanipulation	Damage-free gripping of nano- and micro-sized	MEMS handling tool
construction	Drones and bots for inspections	Window cleaning devices
Prosthetics and orthotics	For assistive devices for user comfort and the security of patients	Linings and support orthoses
Military application	Climbing gear, surveillance at the border or coast, and military operations like surgical strikes, etc.	Soldier-assisted gloves and robots

6.10 Biology of Gecko Feet

The gecko's adhesion is due to unique biological structures found on their feet, perfected through several years of evolution. These structures enable geckos to cling to and climb virtually any surface, smooth wall or ceiling, specifically, without any need for liquids, suction, or chemical adhesives.

6.10.1 Structure of Gecko Feet

The feet of geckos are covered with tiny hair-like structures called setae. Each seta is approximately 0.1 mm long, and along it branches out into hundreds of nanoscale tips called spatulae. The spatulae are usually only about 200 nm wide, making them sufficiently small to engage with surface molecules. The number of spatulae on one gecko toe may run into millions and thus provide a large area for adhesion [28].

6.10.2 Role of Setae and Spatulae

The setae and their spatulae pads are the chief constituents of gecko adhesion. These enable the geckos to take advantage of weak intermolecular forces called van der Waals forces that arise when the spatulae are near a surface. Though weak individually, combined millions of spatulae induce very high adhesion.

6.10.3 The Hierarchical Structure: Setae and Spatulae

The hierarchical organization of structures found on gecko toe pads lies at the heart of adhesion. The macrostructure of a gecko's toe pads features lamellae — flat ridges aligned in parallel lines. These lamellae are lined with thousands of tiny hair-like structures called setae. Each seta branches into hundreds of terminal projections or spatulae that measure just a few nanometers in width, allowing for intimate contact with the surface. The size and packing of these structures are critical for adhesion. Studies show that the molecular contact area of spatulae creates an additional contact area, which greatly increases the adhesive strength due to van der Waals interactions [1]. The alignment and spacing of setae allow geckos to maintain a balance between adhesion and mobility, a critical adaptation for their arboreal and terrestrial lifestyles [7, 10].

6.10.4 Material and Mechanical Properties of Setae

Setae consist mostly of beta-keratin, a protein that offers a blend of stiffness and pliability. This composition is essential for preserving the setae's structural integrity during repetitive stresses and changing environmental conditions. Studies show that these structures can withstand thousands of cycles of attachment and detachment without substantial degradation [13]. The setules of the setae are composed of flaps/bridges allowing for elastic deformation, and they exhibit remarkable compliance, which plays an important role during adhesive contact formation and breakage. The small curvature of setae enhances their conformability on surface defects, which in turn maximizes the contact [6].

6.10.5 Adaptations for Environmental Variability

Geckos exhibit morphological adaptations that allow them to adhere under a range of environmental conditions. Geckos in a moist habitat are a specialty because the van der Waals interactions are supplemented by capillary forces to maintain adhesion [15]. Additionally, research conducted by Autumn *et al.* [2] demonstrated that the hydrophobic properties of the spatulae allow water to be shed, so that it cannot interfere with adhesion in wet environments. Moreover, different gecko species display divergent foot morphologies associated with their ecological roles. For example, arboreal geckos have wider toe pads with more numerous setae, which increase climbing abilities. However, terrestrial species have comparatively narrow toe pads for speed rather than adhesion [3, 9].

6.10.6 Role of Claws in Complementing Adhesion

In addition to the sticky pads, the claws on the gecko's feet also aid in anchorage on rough or irregular surfaces. Claws are used as secondary stabilizing systems, especially when the adhesive forces of the toe pads are inadequate. This dual functionality allows geckos to adapt to an array of surfaces, from smooth leaves to rough tree bark [7].

6.10.7 Evolutionary Perspective on Adhesive Morphology

This evolutionary trajectory illustrates the balance between functional optimization and the ecological requirements of adhesive systems in geckos. Initially, researchers suggested that gecko adhesion probably developed from basic friction mechanisms, gradually evolving into the complex hierarchical configurations present in modern times [14]. This evolutionary trajectory demonstrates the game between morphology and environmental pressures in forming adhesive performance.

6.11 Mechanism of Adhesion

Geckos can stick to and walk on almost any surface owing to their complex mechanisms that act at both the micro- and nanoscales. Unlike conventional mechanisms of adhesion, geckos use a blend of physical forces and structural adaptations, instead of stickiness or suction [29].

6.11.1 Van der Waals Forces

The van der Waals forces represent the primary adhesive force in these animals; they are weak intermolecular, end-range attractive forces that occur between molecules in closest proximity. These forces arise from the temporary dipoles produced due to the movement of electrons around atoms. Although each force is weak in isolation, the combined effect of millions of nanoscale contacts amounts to an enormous adhesive strength, thus allowing the gecko to support its entire body weight on a single toe [29].

6.11.2 Hierarchical Structure of Gecko Feet

The adhesion mechanism in geckos is facilitated by a hierarchical structure that begins with the toes and follows on through to microscopic and then to nanoscopic levels. Geckos have several flexible toes that can splay apart, creating maximal contact with surfaces. In the case of the setae, each toe possesses microscopic hair-like setae that afford it camouflage, branching into hundreds of smaller tips called spatulae. The spatulae, the nanoscale tips about 200 nanometers in width, would come into direct contact with the surface molecules, allowing for intimate contact and maximizing the van der Waals interaction.

6.11.3 Contact Optimization

The spatulae can make intimate contact with various surfaces, ranging from smooth, rough, dry, and wet materials. Their presence allows geckos to adhere to surfaces of different textures and properties [29].

6.11.4 Detachment Mechanism

The gecko can then very rapidly and effortlessly detach one foot from the surface by using the toe itself. In this manner, the spatulae lose intimate contact with the surface, which causes the break in the van der Waals reactions. Rapid attachment and detachment allow the gecko to move swiftly and efficiently.

6.12 Contributions of Friction to Adhesive Performance

Besides van der Waals forces, friction also plays a key role in the attachment of geckos, which becomes especially important during locomotion and under load [14]. Frictional adhesion occurs when shear forces are applied to the gecko's setae. This dual adhesion effect provides stability while climbing and running, as adhesion along the direction of motion is enhanced [18].

Frictional forces are beneficial on slopes. Geckos vary the configuration of their toe orientation to maximize shear force application. This is a strategy that greatly increases adhesive robustness on difficult terrains [18].

6.12.1 Impact of Surface Energy and Wettability

Another important factor affecting gecko adhesion is surface energy, a measure of the interaction potential between a material and its surroundings. Hydrophobic properties: Gecko spatulae remain adhered regardless of humidity or water; this is an additional property that is very valuable for adhesiveness and has already been studied. Geckos will have the greatest performance on mid- to high-energy surfaces [15]. In wet states, diffusion and other capillary forces come into play and alter the energy regimes of surface tension. Geckos use these forces to stick to wet surfaces [2]. The microscopic air pockets stop the water from interfering with the van der Waals interactions, allowing gecko toe pads to cling surprisingly well.

6.12.2 Geometric Contributions to Adhesion

The setae and spatulae are arranged geometrically and finely tuned for adhesive action. The hierarchical structure not only expands the contact area but also allows geckos to modulate the direction and magnitude of their adhesive force dynamically. The spatulae are radially oriented to enable gonads to activate and deactivate adhesion easily, a performance that is vital for locomotion and sticking quickly [5]. The setal curvature has great importance in the adhesive ability of geckos on varying rough surfaces. Their curvature allows the spatulae to contour to irregularities to ensure full contact with the surface [13]. The spacing and orientation of the setae contribute to loading uniformly throughout the setae, which prevents highly localized stresses that could limit adhesion [16].

6.12.3 Interplay of Forces and Structural Features

The adhesive performance of geckos does not rely on any one form of attachment force or structure; it is a result of the combined effect of two influential forces and structural properties [2]. The van der

Waals forces, friction, and surface energy dynamics all contribute in concert to give geckos their impressive adhesive properties [18]. For example, the orientation of setae in shear forces promotes frictional and van der Waals contributions [2]. Furthermore, this interplay allows for the reversible behavior of gecko adhesion. Hierarchical assemblies of the setae and spatulae, alongside material properties, enable geckos to switch between hard adhesion and easy detachment [10].

6.12.4 Influence of Environmental Factors

For example, the mechanisms of adhesion are due to environmental conditions (e.g., temperature, humidity, substrate smoothness). Geckos are extremely adaptable; they modulate their adhesive forces to adapt to environmental changes [15]. In phenotypic changes using three-dimensional models, it was discovered that geckos use greater capillary forces in humid conditions, while using more van der Waals forces in drier areas. Evolutionary modifications have also been implicated in the gecko's ability to adhere to a wide range of surfaces. It is found that arboreal geckos with maximum adhesion on smooth surfaces used more closely spaced setae than geckos with entanglement features, which require speed and reduced adhesive drag to allow the animal to cross at speed [10].

6.12.5 Technological Implications and Future Directions

Research on gecko adhesion systems holds significant promise for the advancement of bio-inspired adhesives and climbing robots. Biomimetic adhesives could have applications from medical devices to space exploration [25]. As Stark *et al.* recommended, future research should focus on the molecular dynamics of van der Waals interactions and the biomechanical properties of setae. Progress in these domains will enable the development of more efficient and scalable adhesive technologies [8].

6.12.6 Directional and Reversible Adhesion

The anisotropic character of gecko adhesion is one of the distinguishing features that enables geckos to stick firmly in one direction and release effortlessly in the opposite. This, along with the reversible nature of their adhesion, allows geckos to traverse a wide range of surfaces with incredible speed and efficiency. The orientable and reversible interactions of gecko adhesion are determined by the structural arrangement of the adhesive apparatus and the physical principles controlling the mechanics of contact.

6.12.7 Anisotropy in Adhesive Forces

The orientation-dependent adhesion of geckos is primarily determined by the asymmetric structure of their setae and spatulae. The configuration of the spatulae at certain orientations allows for an optimal set-up for maximum adhesion and minimum resistance during detachment. The anisotropic nature of force transmission is further accentuated for the hierarchical arrangement of setae set up on lamellae to preferentially give rise to directed force transmission [20]. Geckos modulate toe positioning during use, orienting the setae to maximize engagement. Such controlled adhesion enables geckos to stick securely to surfaces when climbing vertically and switch smoothly between varying angles. Exposing the spatulae at different angles helps them better "hug" surface irregularities. The spatulae are optimally arranged in the direction of movement so that they provide "amplified" adhesion in the direction of movement [18].

6.13 Mechanisms Underpinning Reversibility

The reversibility of gecko adhesion is likely due to an intricate interplay between counteracting adhesion and detachment forces. The van der Waals forces, which govern gecko adhesion, can be adjusted by changing the contact area of the spatulae with the substrate. Rapid detachment in geckos is achieved by decreasing the effective area of contact, often exploiting peeling motions. The detachment of a toe pad at an angle decreases adhesive forces with increased distance. This mechanism enables geckos to detach from substrates with low energetic costs, which is an essential prerequisite for dynamic adhesion [10]. Furthermore, Drotlef *et al.* [12] proposed that setal curvatures were important for detaching because curved structures intrinsically lessen contact under tension.

6.13.1 Functional Role of Load and Shear Forces

This concept is inherently connected to the relationship between the normal and shear forces applied in locomotion. Geckos can produce shear loads in the plane of contact with substantially higher adhesion forces. This behavior aligns the setae and spatulae so they approach the substrate at an angle that maximizes van der Waals interactions. In contrast, shear forces are minimized to facilitate disengagement, exemplifying the temporary aspect of their adhesive ability [6]. Stark and Mitchell [9] extended this idea to propose that the load forces were responsible for the adhesion as the spatulae are engaged evenly across the toe pad. This distribution mitigates stress concentrations that could otherwise impair adhesion. The deformation of the setae under load facilitates their ability to conform to surface irregularities, thereby maintaining reliable adhesive ability.

6.13.2 The Role of Structural Hierarchy in Directionality

A significant aspect of the directional adhesion of the gecko adhesive system is the hierarchy from the macro-, micro-, meso-, and nanoscales. These mechanisms work in tandem to allow geckos to dynamically modulate adhesion *via* lamellae, setae, and spatulae. The hierarchical configuration is said to permit independent yet coordinated engagement of adhesive elements, promoting robust attachment in one direction and a quick detachment in the reverse [2]. Implications of this hierarchy for synthetic adhesives, emphasizing that copying the anisotropic behavior of gecko spatulae would enhance the adhesive potential of the technology as more efficient and versatile. The results highlighted that the geometry orientation of nanoscale structures is necessary for the combination of attachment and removal functions [18].

6.13.3 Environmental Influence on Directional and Reversible Adhesion

Many environmental components directly influence the gecko's adhesion's directional and reversible characteristics, such as substrate roughness, humidity, and temperature. Geckos are highly adaptive, adapting their adhesion mechanisms to their environments. For example, in moist conditions, geckos utilize capillary forces along with van der Waals interactions to maintain adhesion. In the dry situation, on the other hand, the directional adhesion is more dependent on the alignment and orientation of the spatulae [9]. Gecko species occupying divergent ecological niches display morphological differences in their adhesive systems in response to the ecological requirements of their microhabitats. For instance, the toe pads of an arboreal gecko are wider and densely packed with setae for improved adhesion to smooth surfaces, while terrestrial geckos prioritize speed and detachment effectiveness on rough areas [3].

6.13.4 Evolutionary Adaptations for Directionality and Reversibility

The role of evolutionary fine-tuning of unidirectional and reversible adhesion in geckos underscores their ecological adaptability and survivability. These adaptations originate from the most basic friction systems, gradually transforming into the complex anisotropic systems we see today. The evolutionary pressures leading to these functional changes, structurally and behaviorally, are further exemplified in the gecko by attachment adaptation to specific ecological challenges [2].

6.13.5 Applications of Directional and Reversible Adhesion in Biomimetics

Geckos are a popular model organism for bio-inspiration in the fields of locomotion and adhesion, including climbing robots and medical adhesives. Recreating the anisotropic performance of gecko spatulae in synthetic systems would transform adhesive technologies. These improvements could have uses from robotic devices to orthopedic instruments, where sticking and unsticking must be quick [5]. As put forward by Autumn *et al.* [2], future works should focus on studying the molecular dynamics of directional adhesion and the biomechanical attributes of hierarchical structures. These could lead to more high-throughput, scalable approaches to bio-inspired adhesion.

6.13.6 Integration of Mechanisms in Locomotion and Adaptation

The directional and reversible adhesion system of geckos is closely related to gecko gait and habitat adaptability, which has also become a hot topic in the study of geckos. Geckos modulate their adhesive forces during locomotion in a manner that maintains adhesion over a relatively large parameter of space while remaining an energetically efficient process. Control over the orientation, force distribution, and the detachment timing of toe pads results in this modulation [6]. Gecko adhesion is also notable for its ability to accommodate surfaces of different roughness and angles. Geckos change their adhesive strategies depending on the properties of the substrate, demonstrating the versatility of their directional adhesion mechanisms. This plasticity is a major reason for their ecological success and offers lessons for building robust, flexible synthetic systems [9].

6.14 Materials and Design Principles in Gecko-Inspired Adhesives

The art of synthetic adhesives takes great advantage of design principles and materials science. The ultimate goal is to mimic the specific features of gecko adhesion, such as strong and reversible adhesion, compatibility with uneven surfaces, and non-residue application [18].

6.14.1 Material Selection

In attempts to imitate the properties of gecko feet, materials with significant properties are employed. Polymers, such as polydimethylsiloxane, have been frequently used because they possess mechanical flexibility and compatibility with microfabrication. Elastomers are soft, elastic materials that allow them to conform to irregularities of surfaces and remain in intimate contact. Carbon nanotubes are nanoscale materials used to model structures giving a close resemblance to the hierarchical architecture of the gecko setae and spatulae [8]. A single layer of carbon-atom-thick material, graphene, has been examined due to its extremely high adhesive and mechanical structural properties. The hydrogels, when used in the medical field, have been designed with adhesive characteristics that are different compared to those of gecko adhesion, giving them a soft, flexible, and essentially non-toxic range of adhesive materials [8].

6.14.2 Structural Design Principles

The design of synthetic gecko-inspired adhesives focuses on imitating the hierarchical structures of gecko feet. Synthetic adhesives impinge on multi-level structures. The basic structure starts on a microscale with pillars that resemble setae, topped with nanoscale tips that also resemble spatulae. Geckos are modeled to create maximum surface contact definitions so that the van der Waals forces can be well optimized for surface contact. Materials are to be engineered to conform to different surface textures, similar to natural gecko feet, considering flexibility and conformability. Synthetic adhesives often incorporate features that allow one to control the angle at which they stick and come off, mirroring the gecko's toe placement [9].

6.14.3 Key Design Properties

Synthetic adhesives have been developed to mimic geckos' high adhesion strength. The adhesive should be able to support large loads while maintaining a good grip. The material should act like gecko feet, retaining adhesion after multiple uses. Adhesives must not leave any material on the surfaces when detached. Designs may include features that allow the shedding of dirt and debris during use, so they will operate unstably. The material and design should enable scale-up for industrial applications.

6.15 Production Methods

Advanced manufacturing technologies are essential for creating gecko-inspired adhesives. Mainly, microfabrication, photolithography, and soft lithography techniques produce micro- and nanoscale patterns. Electrospinning, to create fibers that resemble gecko setae, is employed to make setal mimics. Also, three-dimensional printing allows for the customization and scale-up of adhesive structures, and the development of carbon nanotubes or graphene-based materials [9].

6.16 Applications of Gecko-Inspired Adhesives

6.16.1 Soft Robotic Gripper with Gecko-Inspired Adhesive

This study uncovers a soft robotic gripper incorporating fluidic elastomer actuators and adhesives inspired by geckos for further improved grasping capabilities. The inspired adhesives, funded after the model of gecko toes, rely on van der Waals forces to supply strong and controllable adhesion, allowing the gripper to handle greater numbers of irregular or fragile objects with enhanced grasping force and efficiency. This gripper demonstrated excellent performance characteristics. It supported loads of more than 111 N under low pressure while remaining lightweight (48.7 g) and low-cost (7.25 dollars). Faster actuation, lower energy consumption, and versatility toward addressing industrial automation, space exploration, and handling complex objects are all highlighted in this innovation, which elucidates the promise of bio-inspired materials in the advancement of soft robotics [8].

6.16.2 Robotics and Automation

Robotic systems have benefited enormously from the development of adhesives after the fashion of geckos that permit wall-climbing robots, even on ceilings and over many difficult-to-reach surfaces. Climbing robots have significant promise for search-and-rescue missions, traversing rubble and reaching even the hardest-to-reach areas. Further, they find utility in erecting and maintaining different infrastructures around tall buildings, bridges, and wind turbines. Relying on gecko adhesion would undoubtedly give robots a strong working grip on various surfaces and strengthen the general operation of these robots for moving along despite environmental resistance [8].

6.16.3 Applications in Robotics and Space Exploration

The gecko adhesion-inspired research has opened new horizons for robotics, as it can now be witnessed in the making and conceptualization of climbing robots, which can walk on vertical and inverted surfaces. Reports show that robotic mechanisms enhanced with synthetic gecko adhesive have improved locomotion in difficult terrains. Robots took advantage of the anisotropic behavior of gecko-inspired adhesives for strong adhesion when climbing and controllable detachment when moving [25].

In the domain of space exploration, adhesion that mimics gecko toe pads has been employed to overcome challenges in gripping and manipulation in microgravity. Gecko adhesives have adhesive properties in various surface conditions [24].

6.16.4 Advancements in Medical Adhesives

Gecko-inspired adhesives have been widely accepted in healthcare for their compatibility with biological issues and lack of residue. They are used in surgical tapes, dressings, and other medical products requiring assurance when stuck to the skin without consequences. The wet adhesion that they can achieve proves valuable in surgical procedures, a field generally hazardous because moisture can interfere with any other standard adhesive. Moreover, the materials put on such a basis are being developed toward non-invasive medical devices for a safe and/or more efficient proposal compared to conventional solutions [11]. Gecko adhesion has attained a position in medical science, as the use of gecko adhesion in the medical field is for tissue bonding. Adhesives that resemble geckos can be useful and bioactive for wound closure. Their property of non-attachment up to the leaking of secondary chemical glues or stitches results in decreasing the tissue damage and promotes healing. Nanoscale surface characteristics are required for achieving strong adherence to biological tissues. This is a crucial prerequisite for surgical applications [21]. Gecko-inspired adhesives have also been showcased in wearable medical equipment. Utilization in sensor patches and drug-delivery systems in which the reversible characteristics of the adhesive allow it to be reapplied and removed multiple times without impairing skin integrity. This shows that gecko-inspired adhesives can serve various medical purposes [22].

6.17 Future Directions in Gecko Adhesion Research and Applications

The microscopic structures and mechanisms responsible for the adhesion of geckos and other climbing animals are interesting areas of research. Some weaknesses and knowledge gaps in the current research could also be useful for future research endeavors.

6.17.1 Expansion of Knowledge of Adhesion Processes at Molecular Dimensions

One of the most promising future research directions is to understand the molecular mechanisms of gecko adhesion. It has been known that van der Waals forces are the main adhesion force, though Autumn *et al.* have emphasized the need to better understand the role played by capillary, electrostatic, and hydrogen bonding [2].

6.17.2 Investigating Novel Features Derived through Evolution in Different Organisms

Comparative studies of geckos with other adhesive organisms in different species may explain the evolution of the diversity of adhesive systems. Russell *et al.* [14] have emphasized the importance of phylogenetic analyses in the aim of reconstructing the evolutionary history and changes in adhesive traits. These studies can show that distant relatives have been subjected to convergent adaptation, as in the case of anoles and tree frogs. Stark *et al.* [8] stressed the importance of including fossil records to reconstruct the probable gecko phylogeny and how their ability to adhere to surfaces evolved with changes in environmental factors. Such findings may lead to new adhesive designs tailored for certain applications, bridging the gap between biological models and synthetic systems.

A critical challenge is to design synthetic adhesives that match the efficiency and versatility of natural gecko adhesion [2, 14]. Progress in microfabrication and nanotechnology is necessary to generate high-fidelity hierarchical structure replicas. Further, the versatility of hybrid materials derives properties from a synthetic polymer and a bio-derived component, and concurrent improvements in adhesion strength and durability. Such computational models could help speed up workflows by simulating how adhesive systems respond to a range of scenarios while minimizing the need for empirical testing. Expanding applications in robotics and wearable technology include the synthesis of gecko-inspired adhesives and their applications in robotics and wearable devices [16]. Introducing anisotropic adhesive properties makes climbing robots adapt to complex surfaces. The use of gecko-inspired adhesives to enhance robotic grippers and manipulators for a secure manipulation of fragile or non-uniform items [24]. Applications of gecko adhesives in sensor patches, prosthetics, and other medical devices could greatly benefit from the utility of the reversible adhesion and skin compatibility.

6.18 Conclusion

The simplicity of the gecko adhesion phenomenon depends on a tightly adjusted composition of van der Waals forces, directional characteristics, frictional interactions, surface energy kinetics, in addition to the hierarchical structure of their adhesive systems. This versatility of natural systems allows them to adapt to environmental variability, including humidity, temperature, and surface roughness. Some of the most exciting advancements in that respect have not only propelled our understanding of biological adhesion forward but have also deepened our appreciation of how geckos modulate their adhesion to maintain mobility while exerting stability, enabling the design approach to synthetic adhesives, robotic grippers, and biomedical devices. And while technological improvements have tried to mimic these biological characteristics, they still face challenges, such as environmental resiliency, scaling of synthetic systems, long-term durability, and reversibility. Non-invasive biology (morphologies, evolutionary trajectories, molecular interactions) will surely steer innovation into the future. To overcome these challenges and extend the applicability of bio-inspired adhesion technologies, the integration of smart responsive materials and advanced manufacturing techniques will be instrumental.

This is a reminder of how drawing knowledge from different disciplines—biology, physics, nanotech, and materials science—can drive technological advances with real-world impact, as is the case with gecko adhesion. In short, the mechanics of gecko stickiness are anything but academic curiosity. It is a direct lesson from nature, where we learn from nature to create designs that otherwise could not be made. As knowledge is gained and technology advances, gecko-inspired systems will continue to lead the way—providing sustainable, efficient, and adaptable solutions to some of the most pressing challenges facing the modern world, from vertical and upside-down surfaces.

References

1. Autumn, K., Liang, Y.A., Hsieh, S.T., Zesch, W., Chan, W.P., Kenny, T.W., Fearing, R., Full, R.J., Adhesive force of a single gecko foot-hair. *Nature*, 405, 6787, 681–685, 2000. DOI: 10.1038/35015073.
2. Autumn, K., Sitti, M., Peattie, A., Hansen, W.R., Israelachvili, J., Full, R.J., Evidence for van der Waals adhesion in gecko setae. *Proc. Natl. Acad. Sci. USA*, 99, 19, 12252–12256, 2002. DOI: 10.1073/pnas.192252799.
3. Russell, A.P. and Gamble, T., Evolution of the Gekkotan Adhesive System: Does Digit Anatomy Point to One or More Origins? *Integr. Comp. Biol.*, 59, 1, 131–147, 2019. https://doi.org/10.1093/icb/icz006.
4. Bartlett, M.D., Croll, A.B., King, D.R., Paret, B.M., Irschick, D.J., Crosby, A.J., Looking Beyond Fibrillar Features to Scale Gecko-Like Adhesion. *Adv. Mater.*, 24, 8, 1078–1083, 2012b. https://doi.org/10.1002/adma.201104191.
5. Santos, D., Spenko, M., Parness, A., Kim, S., Cutkosky, M., Directional adhesion for climbing: theoretical and practical considerations. *J. Adhes. Sci. Technol.*, 21, 12–13, 1317–1341, 2007. https://doi.org/10.1163/156856107782328399.
6. Higham, T.E., Russell, A.P., Niewiarowski, P.H., Wright, A., Speck, T., The Ecomechanics of Gecko Adhesion: Natural Surface Topography, Evolution, and Biomimetics. *Integr. Comp. Biol.*, 59, 1, 148–167, 2019. https://doi.org/10.1093/icb/icz013.
7. Naylor, E.R. and Higham, T.E., Attachment Beyond the Adhesive System: The Contribution of Claws to Gecko Clinging and Locomotion. *Integr. Comp. Biol.*, 59, 1, 168–181, 2019b. https://doi.org/10.1093/icb/icz027.
8. Stark, A.Y., McClung, B., Niewiarowski, P.H., Dhinojwala, A., Reduction of Water Surface Tension Significantly Impacts Gecko

Adhesion Underwater. *Integr. Comp. Biol.*, 54, 6, 1026–1033, 2014. https://doi.org/10.1093/icb/icu066.
9. Stark, A.Y. and Mitchell, C.T., Stick or Slip: Adhesive Performance of Geckos and Gecko-Inspired Synthetics in Wet Environments. *Integr. Comp. Biol.*, 59, 1, 214–226, 2019. https://doi.org/10.1093/icb/icz008.
10. Garner, A.M., Wilson, M.C., Russell, A.P., Dhinojwala, A., Niewiarowski, P.H., Going Out on a Limb: How Investigation of the Anoline Adhesive System Can Enhance Our Understanding of Fibrillar Adhesion. *Integr. Comp. Biol.*, 59, 1, 61–69, 2019. https://doi.org/10.1093/icb/icz012.
11. Niewiarowski, P.H., Dhinojwala, A., Garner, A.M., A Physical Model Approach to Gecko Adhesion Opportunity and Constraint: How Rough Could It Be? *Integr. Comp. Biol.*, 59, 1, 203–213, 2019. https://doi.org/10.1093/icb/icz029.
12. Drotlef, D.-., Dayan, C.B., Sitti, M., Bio-inspired Composite Microfibers for Strong and Reversible Adhesion on Smooth Surfaces. *Integr. Comp. Biol.*, 59, 1, 227–235, 2019. https://doi.org/10.1093/icb/icz009.
13. Imburgia, M.J., Kuo, C.-., Briggs, D.R., Irschick, D.J., Crosby, A.J., Effects of Digit Orientation on Gecko Adhesive Force Capacity: Synthetic and Behavioral Studies. *Integr. Comp. Biol.*, 59, 1, 182–192, 2019. https://doi.org/10.1093/icb/icz024.
14. Russell, A.P., Stark, A.Y., Higham, T.E., The Integrative Biology of Gecko Adhesion: Historical Review, Current Understanding, and Grand Challenges. *Integr. Comp. Biol.*, 59, 1, 101–116, 2019. https://doi.org/10.1093/icb/icz032.
15. Bauer, A.M., Gecko Adhesion in Space and Time: A Phylogenetic Perspective on the Scansorial Success Story. *Integr. Comp. Biol.*, 59, 1, 117–130, 2019. https://doi.org/10.1093/icb/icz020.
16. Murphy, M.P., Kim, S., Sitti, M., Enhanced Adhesion by Gecko-Inspired Hierarchical Fibrillar Adhesives. *ACS Appl. Mater. Interfaces*, 1, 4, 849–855, 2009. https://doi.org/10.1021/am8002439.
17. Kim, S. and Sitti, M., Biologically inspired polymer microfibers with spatulate tips as repeatable fibrillar adhesives. *Appl. Phys. Lett.*, 89, 26, 2006. https://doi.org/10.1063/1.2424442.
18. Autumn, K., Dittmore, A., Santos, D., Spenko, M., Cutkosky, M., Frictional adhesion: a new angle on gecko attachment. *J. Exp. Biol.*, 209, 18, 3569–3579, 2006b. https://doi.org/10.1242/jeb.02486.

19. Autumn, K. and Gravish, N., Gecko adhesion: evolutionary nanotechnology. *Philos. Trans. R. Soc. Math. Phys. Eng. Sci.*, 366, 1870, 1575–1590, 2008b. https://doi.org/10.1098/rsta.2007.2173.
20. Parness, A., Soto, D., Esparza, N., Gravish, N., Wilkinson, M., Autumn, K., Cutkosky, M., A microfabricated wedge-shaped adhesive array displaying gecko-like dynamic adhesion, directionality and long lifetime. *J. R. Soc. Interface*, 6, 41, 1223–1232, 2009b. https://doi.org/10.1098/rsif.2009.0048.
21. Mahdavi, A., Ferreira, L., Sundback, C., Nichol, J.W., Chan, E.P., Carter, D.J.D., Bettinger, C.J., Patanavanich, S., Chignozha, L., Ben-Joseph, E., Galakatos, A., Pryor, H., Pomerantseva, I., Masiakos, P.T., Faquin, W., Zumbuehl, A., Hong, S., Borenstein, J., Vacanti, J., Karp, J.M., A biodegradable and biocompatible gecko-inspired tissue adhesive. *Proc. Natl. Acad. Sci.*, 105, 7, 2307–2312, 2008. https://doi.org/10.1073/pnas.0712117105.
22. Duarte, A., Coelho, J., Bordado, J., Cidade, M., Gil, M., Surgical adhesives: Systematic review of the main types and development forecast. *Prog. Polym. Sci.*, 37, 8, 1031–1050, 2011. https://doi.org/10.1016/j.progpolymsci.2011.12.003.
23. Huber, G., Mantz, H., Spolenak, R., Mecke, K., Jacobs, K., Gorb, S.N., Arzt, E., Evidence for capillarity contributions to gecko adhesion from single spatula nanomechanical measurements. *Proc. Natl. Acad. Sci.*, 102, 45, 16293–16296, 2005. https://doi.org/10.1073/pnas.0506328102.
24. Glick, P., Suresh, S.A., Ruffatto, D., Cutkosky, M., Tolley, M.T., Parness, A., , A Soft Robotic Gripper with Gecko-Inspired Adhesive. *IEEE Rob. Autom. Lett.*, 3, 2, 903–910, 2018. https://doi.org/10.1109/lra.2018.2792688.
25. Menon, C., Murphy, M., Sitti, M., Gecko Inspired Surface Climbing Robots, in: *2004 IEEE International Conference on Robotics and Biomimetics, Shenyang, China*, pp. 431–436, 2004, Doi: 10.1109/ROBIO.2004.1521817.
26. Wang, W., Liu, Y., Xie, Z., GeckoLike Dry Adhesive Surfaces and Their Applications: A Review. *J. Bionic Eng.*, 18, 8, 1–34, 2021.

27. Boesel, L.F., Greiner, C., Arzt, E., del Campo, A., Gecko-Inspired Surfaces: A Path to Strong and Reversible Dry Adhesives. *Adv. Mater.*, 22, 19, 2125–37, 2010.
28. Pianka, E.R. and Sweet, S.S., Integrative biology of sticky feet in geckos. *Bioessays*, 27, 6, 647–52, 2005.
29. Autumn, K., How Gecko Toes Stick. *Am. Sci.*, 94, 2, 2006.

7

Manufacturing Shark Skin-Inspired Swimwear with Enhanced Fluid Dynamics and Performance

Abstract

Shark skin has been one of the most researched biomimetic technologies of the last decade by various researchers for its capability to reduce drag due to the presence of denticles and ridges in a specific pattern. These denticles are also known for their anti-biofouling properties, as well as friction reduction properties. This paper is a detailed study based on several research papers about the latest technologies based on recent studies on sharkskin properties. This paper provides an overview of various properties of shark skin, considering various shark skin-based fabrication techniques with extensive focus on the mechanical properties of denticles on shark skin. It also discusses their role in enhancing the aerodynamic properties during motion. This paper also provides some insights into future research techniques to be implemented, such as 3D printing of shark skin-based swimwear with emphasis on the bristling effect.

***Keywords*:** Shark skin, denticles, fabrication techniques, aerodynamic properties

7.1 Introduction

Sharks are one of the most interesting animals that have lived on Earth for millions of years through various evolutionary techniques. Their skin evidently is the most fascinating feature and is studied extensively for small, toothlike structures present on it called denticles. These are made of an enamel-like substance and contain grooves or patterns often called ridges. According to their position on the body, these denticles have several dimensions and forms, arranged in an overlapping pattern, typically having a width of 0.1mm to 1mm [1]. The denticles of sharks, combined with their

Jagadeesha T. and Sandip Kunar. Bio-Inspired Manufacturing Systems, (163–184) © 2026 Scrivener Publishing LLC

surface ridges, reduce the drag, increasing swimming efficiency. The ridges on each denticle are aligned with the flow of water, forming a small, patterned surface, eliminating the formation of turbulent eddies and allowing water to flow smoothly along the shark's body. The ridges also form tiny grooves that reduce the interaction between the skin and the surrounding water. This phenomenon is known as the riblet effect, where the turbulence of the flow is concentrated at a specific place, thereby ensuring smooth laminar flow over the surface. These denticles also have a bristle-like structure where these small patterns protrude and stand out, creating a sharp point on the surface of the shark; therefore, decreasing the boundary layer effect and drag associated with water flow. These bristles also help with the anti-biofouling effect by creating layers one over the other, thereby reducing the ability of microorganisms to stick to the surface. Several researchers conducted a study and simulated turbulent flow over surfaces with various riblet structures. They found that riblets help in the formation and motion of streamlined vortices, increase the thickness of the viscous sublayer, resulting in a lubrication-like effect and decreasing friction over the surface. Furthermore, they also reduce flow fluctuations near the surface and lower stress by ensuring a smooth flow near the surface [2]. Studies also show that these riblet structures, maintaining a low h, h/s, t, and t/s ratios where h is the height, t is the thickness, and s is the spacing between the riblets, reduce the drag up to 34% in water and about 26% in air [3]. Further studies also indicated that denticles not only reduce drag but also affect the behavior of vortex formation, particularly on the tail and pectoral fins, where they can increase thrust and lift for speed and control [4]. Bio-inspired designs based on shark skin have been investigated for their ability to boost hydrophobic characteristics and minimize hydrodynamic drag. Experimental results suggested that original shark skin surfaces achieve drag reduction efficiencies 12% more compared to smooth surfaces, with stretched patterns increasing these properties even further [5]. Some studies also focused more on lift properties using denticles. One such study found that the denticle profiles enhance performance by causing airflow to reattach smoothly after separation, increasing suction along the chord and lift. The spanwise curvature improves the overall performance by maintaining lift and ensuring smooth fluid movement across the surface [6].

By incorporating advanced scanning technology and polymer composites, new methods for replicating natural surfaces can be developed. This allows better performance in applications, such as filtration and surface protection for making fabrics inspired by shark skin [2]. In 2018, various fabrication methods, such as bio-replication and bio-scaling, were developed to replicate riblet structures for wider applications, including antifouling [7]. A 2014 study focused on designing a drag-reducing material by merging biomimetic shark skin structures with advanced polymers. Fourier transform infrared spectroscopy provided detailed insights into the material's efficiency and structure. The optimization of polymer additive diffusion ports was conducted to enhance the drag reduction rate while minimizing additive waste. The results generated were able to decrease drag by 80%, highlighting an effective combination for the shark skin pattern and polymer additives in achieving significant drag reduction [8].

Recent studies in 2024 highlighted several biomimetic shark skin materials, but these materials failed to replicate the exact dimensions and distinct parallel ridges of shark skin. Most research conducted on riblet structures is primarily on rigid surfaces, and not on moving skin surfaces. Advanced 3D profilometry enabled the collection of more accurate data, but current manufacturing techniques cannot replicate these surfaces accurately. Moreover, the study also suggested that future research must be based on the effects of denticle texture on drag and lift forces [2]. This chapter aims to provide information on the current state of studying shark skin and its recent trends in manufacturing biomimetic shark skin designs.

7.2 Morphological, Functional, and Hydrodynamic Properties of Shark Skin

The shark skin has denticles in a layered structure. As part of several research studies, it has been found that denticles are made up of enamel and dentine, which are present in teeth. The riblets on the denticles are longitudinally oriented in the direction of water flow and are stacked in rows to reduce skin resistance. Riblets are often present on the leading surface of denticles, showing variations in their numbers and dimensions depending on the position and species of shark. A 3D model shows that the central ridges are taller than the lateral ones, although this depends on the species.

As part of research, it was found that denticles of the small-spotted catshark had a length of around a few hundred micrometers to about a millimeter. A representative picture is shown in Figure 7.1. with denticles having a smaller size concentrated near fins and gills, while larger ones are present on the body. For the shortfin mako shark, the denticle lengths are between 105 ± 7 μm and 286 ± 25 μm. Widths are almost the same as the lengths of the denticles, and with a length-to-width ratio of 1 to 1.5. Unlike other bony fish, shark denticles do not grow as the shark matures; their size and structure depend on species [1].

Several observations are made regarding the lattice structure and the shapes of denticles within the species, as well as within different species. The shortfin mako shark (*Isurus oxyrinchus*) has a varied density of denticles along both the horizontal and perpendicular planes. Higher densities are observed in the upper and lower surfaces and the front section of the body, with the lowest concentration

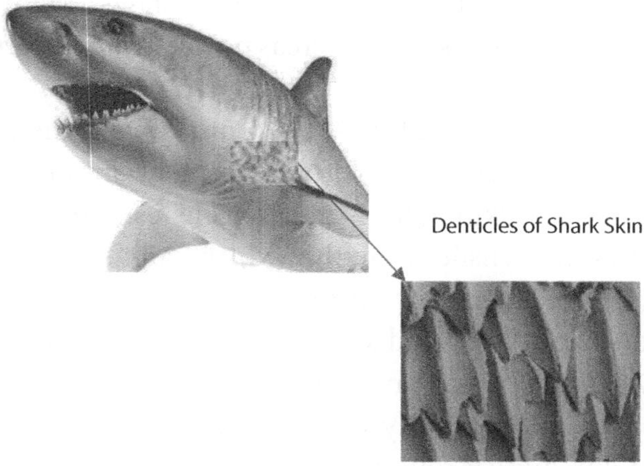

Figure 7.1 Denticles of shark skin.

along the lengthwise trajectory. Density increases again toward the posterior region.

Shark skin is also known for its peculiar ability of anti-biofouling. This is possible because of the layered structure of denticles, creating turbulence at the boundary, allowing small eddies to deter these organisms. Furthermore, due to the rough texture of the skin, air is trapped inside, making the skin hydrophobic. From research carried out in 2003, juvenile dogfish exhibit a behavior called "scale rasping." This shark uses backward-facing denticles on its skin as natural hooks to grip and hold prey. By attaching food to the skin, specifically along the sides and tail region, the shark effectively keeps the prey in place.

In 1997, the material properties of shark skin, focusing on its stress–strain characteristics under tension, were observed. The study revealed differences in the skin's mechanical properties based on orientation and regions. Shark skin samples were tested in four directions (longitudinal, perpendicular, and two diagonal orientations) from the anterior and posterior regions. Results showed that longitudinal samples exhibited higher strain rates, while posterior samples demonstrated greater stiffness. This difference is due to the wavy structure of fibers in the skin, which requires force to straighten before fully resisting stress. Anterior fibers show a significant 120° deviation from the stress axis, whereas posterior fibers were deviated by only 15°.

Research was conducted in 2018 to understand the relationship between the influence of denticle arrangement on the physical performance of shark skin. Quasi-static tensile tests showed variations in properties like tensile strength, with the highest ultimate tensile strength (UTS) in the top-front zone (D1) or tail (L3), depending on the species. The bonnethead shark exhibited the maximum value of UTS at 47 MPa in the tail and the lowest at 9 MPa in the abdominal side. The maximum rigidity is found in the top-front zone, while maximum durability is found in the species, scalloped hammerhead and blacktip sharks. While UTS is not significantly affected by the density of the denticles, it was directly proportional to stiffness and inversely with toughness. Research in 2008 on shark scale bristling has revealed that shortfin mako sharks can adjust their scales dynamically during high-speed swimming, increasing hydrodynamic efficiency. This capability is attributed to the stratum compactum, a layered collagen fiber structure forming their exoskeleton, which balances longitudinal and hoop stresses while storing elastic energy for efficient movement. Later in 2012, it was found that mako shark scales were smaller, slimmer, and had fewer ribs compared to those of blacktip sharks. Highly flexible triangular scales extend from the gill region to the tail, allowing even a 50° rotation after their death.

In 2018, a thorough examination focused on the denticle structures at 20 distinct points on the body of the dusky smoothhound shark (*Mustelus canis*). The research focused on metrological variables, such as roughness (Sq) for skin samples. The research included measurements of denticle length and width for each region, along with the spacing between adjacent ridges and the height of specific denticles. The longest and widest denticles were found at the tip of the shark's nose, while the shortest and narrowest denticles were located just behind the upper fins. Roughness values were highest towards the upper part of the body and lowest towards the rear. A computational fluid dynamics (CFD) analysis conducted further indicated how the non-uniform height and spacing of these grooves impact turbulent flow over the skin. Results showed that increasing the height and spacing ratios of the grooves increases the effectiveness of secondary flows. This highlights the hydrodynamic advantage of irregular grooves in improving the natural flow of water near the shark skin. Further studies conducted in 2018 confirm the importance of non-uniform and multiple ridged grooves in controlling the movement of water across the denticles. The shortfin mako shark can control the movement of water over the surface of its skin, increasing its swimming efficiency. Studies conducted in 2021 on the mako shark showed that the bristling angles of the mako shark's denticles range from 30° to 50°, where 50° is the most effective angle for the smooth water flow and higher swimming efficiencies.

High-resolution imaging techniques have been widely used to study the morphology of shark denticles. This allows for an accurate 3D mapping of the denticle structures for better 3D analysis. Particle image velocimetry (PIV) and CFD are two widely used methods to analyze the hydrodynamic properties of shark skin. The technique PIV is for optical measurement of fluid flow, which can analyze the pattern and velocity within fluid flow to understand how fluid behaves around the denticles. On the other hand, CFD simulates and analyzes the behavior of flow interacting with the denticles under various conditions.

7.3 Fabrication, Applications, and Hydrodynamic Properties of Biomimetic Shark Skin

Research on replicating shark skin is primarily focused on making simplified patterns like riblets and replicating shark skin using advanced techniques, such as 3D printing. In 2016, research had indicated three primary methods among all the methods for manufacturing biomimetic shark skin: bio-replicated forming, direct fabrication, and indirect fabrication of detailed surface features. The bio-replicated forming method uses actual shark skin as a template. A polymer is poured over the surface of the species to form a mold;

it is used to make a cast to make a highly accurate model. While this technique gives a very precise surface replicating natural shark skin, it is not advised due to their high costs and resource constraints. In the direct fabrication method, microstructures are made directly without relying on a template, using techniques like surface machining to create biomimetic samples. While this gives precise models, it is limited by due efficiency of the methods involved. In the conventional manufacturing method, grinding is used to create a mold. However, unlike the bio-replicated forming method, the mold is directly made without casting a polymer over a biological template.

In 2016, several studies used 3D printing to create biomimetic skin over traditional methods, such as milling or casting. The speed of 3D printing helps in the rapid production of large artificial structures and makes structures with varied mechanical properties, such as integrating rigid, shark-inspired denticles into flexible sheets. It can also help in the precise control of denticle properties, such as size, spacing, patterns, and mechanical properties. However, replicating the true structural properties of real shark skin is still a significant challenge, as these features cannot be achieved either by casting or simple mechanical fabrication methods. Later in 2021, photo reconfiguration was introduced, a unique method for crafting intricate, bio-inspired structures at the micrometer level. This method has two stages: (1) Creating a pattern of small columns with a mold, and (2) photo-reconfiguring these small columned patterns to change their shapes. A soft casting mold is used to shape an azopolymer, forming a base array of patterns. These small patterns are then modified through exposure to light directed at an angle, utilizing interference patterns from two circularly polarized sources. This type of lighting causes the formation of asymmetrical riblet patterns on the upper layer of the structure.

In 2023, the influence of denticles-like structures on shark skin-inspired fabrics was identified. Research was done on the physical characteristics of shark skin material with varying sizes of particles, giving significance to strength under stress, the elasticity modulus, and elongation until failure. Using a 3D printer with high-resolution photopolymerization capabilities, intricate shark skin fabric microstructures are produced from flexible 80A resin.

This research showed a significant relationship between denticle size and mechanical performance. In the lateral direction, fabrics with smaller particles have the highest resistance to tension (8.73 MPa), but increasing particle size by 50% resulted in a 29.67% reduction (6.14 MPa). A 20% increase in size caused an 18.35% decrease (from 7.52 MPa to 6.14 MPa). Fabrics with medium-sized denticles showed the maximum tensile strength and breaking elongation when measured along the length. Compared to small denticles, medium denticles increased elongation at break by 7.17%, while large denticles reduced it by 19.07%. In the lateral direction, small denticle fabrics also exhibited superior elongation, showing their ability to enhance flexibility in this orientation.

Several studies have been done to simplify the structures and shapes of denticles into cost-effective forms, making them useful for practical applications. For example, in the aircraft industry, having riblets on 70% of the surface area reduced fuel consumption by 3%, while in pipe flow engineering, surfaces with riblet structures could have a resistance reduction of at least 5%. Among various riblet geometries, blade-type riblets show the highest swimming efficiencies by reduction of drag up to 9.9%, as compared to scalloped and sawtooth types. In 2018, three variations of shark denticle-inspired designs on air foils were tested. The first mimicked natural shark denticles, the second had a simple 2D bump, and the third had both structures, forming a continuous shark-inspired profile. Different angles of attack were tested, and the continuous shark-inspired profile had better properties than the other designs. It showed a higher lift-to-drag ratio and lowered drag at larger angles, and has better lift even at low angles. This profile was not only aerodynamically superior but also could replicate complex structures of real-life shark skin patterns, making it useful for industrial manufacturing.

These findings highlight the ability to produce continuous shark skin patterns for various applications, such as wind power blades, UAVs (unmanned aerial vehicles), and autonomous water vehicles.

The mechanisms might also offer significant advantages at higher flows, increasing the air and water mobility. In 2010, it was found that vortices above riblets lead to localized high-shear stresses when interacting with the tips, while the valleys experience significantly lower shear stresses. This distribution reduces the drag caused by the enhanced area. Although some auxiliary flow disturbances are present in the riblet channels, their contribution only slightly increases shear stress. In 2018, further investigations proved that non-uniform grooves are better than uniform grooves at enhancing secondary flows.

Later again in 2018, research was conducted on numerical analyses of biomimetic shark skin-inspired modifications for rectangular and container ship models. The results indicated reductions in stress along the wall, with a 3.75% decrease in the coefficient of drag and a 3.89% reduction in drag force for the container ship model. However, further study is needed on experimental validation to fully realize the benefits and broaden the practical applications of these innovations. Several studies analyzed the hydrodynamic abilities of shark skin-inspired fabrics and structures. A study in 2012 showed the effectiveness of shark skin surface in increasing self-propelled swimming speed using a robotic flapping foil device. For precise measurements of speeds and control flow dynamics in the device, PIV is used. Both rigid and flexible sharkskin foils, along with two biomimetic shark skin models, were tested to assess their impact on propulsion. Results showed that the curvature values of flexible shark skin foils (0.17–0.25 cm^{-1}) closely matched the mid-body curvature values observed in live spiny dogfish swimming in laboratory flow tanks. This suggests that shark skin membranes replicate the natural bending behavior of live shark skin.

The findings from the 2012 study also reveal that denticles not only reduce drag but also help change the vortex location, particularly on the tail surface, increasing thrust, improving lift, and maneuvering forces on pectoral fins. This suggests a broader functional role for denticles in hydrodynamic performance. Studies from 2014 suggest that biomimetic shark skin can reduce static drag by 8.7% at low speeds and increase swimming speed by 6.6% while reducing energy consumption by up to 5.9%. The study emphasized the formation of improved leading-edge vortices, which enhance swimming dynamics. Later in 2018, researchers examined denticle size, finding that smaller denticles optimized swimming performance and energy efficiency, particularly at higher speeds. However, challenges while replicating the actual denticles limited the accuracy of the synthetic models. Researchers also investigated the bristling effect of mako shark skin using magnified oil-channel experiments. Although closely spaced scales reduced shear stress, they were less effective than optimized ribbed surfaces for drag reduction.

7.4 Manufacturing Techniques of Shark Skin-Patterned Fabrics

There are several methods for manufacturing sharkskin-inspired fabrics, especially using methods mentioned above and digitally. Several studies are first carried out using 3D modeling methods, such as computer-aided design (CAD) software, and then analyzed using finite element analysis (FEA) simulations to design a simulation of fluid flow to know lift, drag, and efficiencies while swimming with these fabrics.

7.4.1 Digital Modeling and Computer-Aided Design

Digital modeling and CAD play a major part in the digital manufacturing of sharkskin fabrics. The first step is to create a 3D model of shark skin patterns, including the ridges, grooves, and microstructural patterns as dermal denticles. Using CAD software, several 3D models with high precision can be made and used for analysis. All the details of curvature of skin must be shown in final drawing. Many sophisticated CAD and computer-aided manufacturing software are available to model in a very systematic way, considering all the complexities and multiple physics. Detailed production drawings facilitate the easy manufacture with minimal reuse and rejection.

All the minutes and sectional views can be shown in CAD software. There is a greater room for visualization of each part, as well as assembly. Conventionally, a lot of time is lost in imaging an assembly of the product, but due to advancements in CAD, all the part drawing and assembly can be visualized with different colors. Almost many collisions can be prevented by proper visualization using CAD software. Both textures and structures can be visualized using software. Shape, size, and behavior under different environmental conditions can be modeled and simulated in custom-made software. The self-cleaning ability can be simulated to reduce drag forces on the skin for enhanced swimming efficiency. Very precise and accurate modeling can be developed considering the multiphysics involved in the analysis of the structure [9].

Finite element analysis can be used to simulate skin structure, and the stress and deformation can be visualized and calculated using the software. Several parameters can be changed, and the exact behavior can be studied before manufacturing. Topology optimization and process parameter optimization can be done to increase flexibility and efficiency under various testing conditions. Finite element analysis can be used to study the stress and strain developed inside the fiber or fabrics, which resembles shark skin. By varying different parameters, various designs can be arrived at, and an optimum design based on least drag and highest efficiency can be chosen. However, manufacturability, ease of assembly, ease of testing and cost must be worked out before finalizing the design. Design may sometimes be optimum with the best efficiency, but may not be cost-effective [10]. Careful analysis, cons, and pros must be studied beforehand. Some environmental conditions must be considered while performing numerical analysis, especially for applications like marine applications, drones, and turbines.

Materials selection is as important as design itself. One advantage of FEA is that the model can be updated with available material data that is obtained through experimentation, and then the behavior of the system or process can be studied by changing the material properties. For example, material with constant thermal conductivity can be modeled first; afterwards, the variation of thermal conductivity with temperature can be updated in the FEA model, and the behavior can be studied. With available options, certain conditions can also be studied for a particular range without conducting experiments. For example, rigidity, flexibility, and durability tests are conducted for particular class of polymer, and data is acquired experimentally. Now, if the behavior to be studied for a range of load values, temperature range, or a set of humidity values can be done or simulated in the FEA package, and optimization is also possible using various optimization techniques. The advanced FEA package allows the modeling of micro-scale textured surfaces, denticles, riblets, and CFD study/simulations can be carried out, and hydrodynamic behavior can be studied using an optimal orientation of fibers in the composite structure, sizing, and position can be modeled. Also, some defects, like the void and delamination in skin structure, can be addressed, and advanced failure theory can be incorporated using the user-defined subroutines.

Finite element analysis can also be used to study the behavior of the skin fabrics under thermal loading and fatigue loading. Skin should retain its mechanical, thermal properties under various loading and temperature conditions. Coupled field analysis studies the thermal stress developed due to the load. Similarly, skin is subjected to various components of stress depending on the orientation. In some cases, or orientations, stresses may be maximum while in other directions, stresses may be zero. The optimum conditions must be obtained by studying the stress direction, and this factor of safety can be arrived at after considering various criteria available for fatigue analysis, such as Soderberg and safer Soderberg relations.

7.4.2 Use of Additive Manufacturing for the Fabrication of Skin

Additive manufacturing is now commonly used to produce complex shapes. Here, material consumption is less compared to other subtractive processes. There are many advanced laser processes used in additive manufacturing processes, including bioprinting [11]. This work involves building the product by layer-by-layer deposition, using a laser or a heated nozzle. Textured surfaces of the shark skin with all details like grooves, stripes, ridges, valleys, and other complex curves can be created using additive manufacturing. Since the material is deposited layer by layer, its precision is quite high, and wastage is minimal. The biggest disadvantage is that they are produced one after the other and are not suitable for mass production. Compared to conventional machining, which uses general-purpose machines, special-purpose machines have a lower efficiency and are also time-consuming.

In conventional machining, machining rate and characteristics depend on the type of tool used, the hardness of the workpiece, and the hardness of the tool. But in additive material, most of the material can be 3D printed. Also, very complex shapes like shark skin can be easily duplicated compared to conventional machining. The complexity of the component that can be produced by conventional machining is limited compared to 3D printing, which is very versatile in terms of materials. Various fabrics, which are inspired by shark skin, can be manufactured from a wide range of materials available in the market, such as thermoplastic elastomers, polyurethane, and polylactic acid. Many materials can be combined during the process of manufacturing of filament. Various properties like flexibility, durability, thermal, mechanical and other properties can be improved by using proper materials. Various biocompatible or medical polymer materials are used if they are to be used for clinical purposes.

Surface topology can be easily optimized using 3D printing. Factors, such as denticle size and shape, can be changed to manufacture different fabrics. A single fabric can be designed into various fabrics using this method of preparation. In sportswear, where the performance of fabrics matters, this method of manufacturing can help maintain the properties even at the time of mass production. Another method, 3D printing, also enables manufacturers to produce fabrics with multiple materials used at once to increase the performance characteristics and comfort of the fabric.

7.4.3 Nano- and Microstructuring

In photolithography, a pattern with essential design is coated with substances that respond to light exposure, and light is projected onto the surface. Thus, the marked designs remain permanently on the surface. This method is exclusively used to make highly complicated structures in shark skin on flexible materials.

Laser ablation is the process of using laser beams with high intensity to remove materials from hard surfaces, making it possible to create patterned surfaces at micro- and nanoscales on brittle, as well as ductile, surfaces. The focus of the laser beam can be controlled as part of controlling the intensity, which allows for the essential trajectory of the laser to create intricate designs on the surfaces. These surfaces can be made to the exact dimensions of patterns present on shark skin, and thus required properties can be maintained.

In electrospinning, micro- and nano-fibers are formed by applying high-intensity currents to a polymer solution. These fibers are then made to exit through a small diametric hole under high pressure. Later, these fibers are molded into a group and are made into different patterns to make shark skin topography. In nanoimprint lithography, a mold with minute patterns (nanometric scale) is pressed onto the substrate surface, and these nanometric patterns are imprinted on the given surface. Due to its precision and accuracy, this method is also used for other complicated applications, such as the manufacturing of micro-electro-mechanical systems. This method also produces surfaces with high consistency and requires relatively less time than many other industrial processes to make nanometric-scale textures on surfaces.

7.4.4 Integration with Flexible Substrates

Certain materials like polyurethane, TPE, can have properties like high strength and water resistance, which can be used as a perfect choice for making fabrics used in swimwear and other water-related applications. These materials are often made into clothes with high levels of surface engineering (micro- to nanometric scales).

Silicone rubber shows high durability and resists changes in structures even under high-temperature applications. Silicone has a smooth touch and thus has the property to repel water from sticking to the surface. This helps maintain water resistance and also helps in preventing friction losses, increasing efficiency while reducing drag. Weaving and knitting different materials together, especially nylon or spandex, which are often used in sports wear applications with microstructural patterns embedded onto the surface, makes clothing extremely flexible, resists drag, and provides us with extreme comfort due to the presence of nanometric holes through which breathability increases.

7.4.5 Coating Techniques

These techniques are often used to coat different materials on substrate surfaces to change the already existing microstructural patterns and the properties of the substrate surface. Spray coatings are a process of spraying air containing micro- and nanoparticles onto the substrate surface, forming a layer of these particles that protects the surface. In this application, hydrophobic materials can be made into nano-scaled particles and then sprayed onto different surfaces to create shark skin replicas. This method provides a cheap and efficient solution to create riblets on the substrate surfaces. Dip coating is used when precise control over the thickness is required. In this process, the substrate material is submerged in the solution of the material to be coated. Later, the substrate is taken out and dried using different heat treatment methods to provide a layer of coating on the substrate surface. A variety of materials, such as metals, textiles, and polymers, can be coated using this technique.

Spin coating is used when a surface needs to be provided with a uniform thickness of coating along the surface to be coated. It involves high rotation of surfaces coated with high-viscosity liquids, ensuring a smooth and uniform transfer of coatings in all directions on the substrate surface. Thin layers of coating are applied several times to make uniform coatings with the desired thickness. Generally, this process is often combined with laser ablation for the micro-structuring of the surface after coating. Nanoparticles, such as silica or titanium dioxide, help improve hydrophobicity and reduce drag, thereby increasing the efficiency. These coatings also provide the substrate with additional protection against substrate oxidation, preventing unwanted rusting. This method is often used in ship hull preparations.

7.5 Sensors and Smart Sensors Used in Sharkskin Fabrics

Sensors and smart sensors are often used in various scenarios where real-time data interfacing is required to increase the performance characteristics. In the case of sharkskin fabrics, these sensors can provide real-time data when testing the surface for various analyses like drag and hydrophobicity, and help in improving the efficiency. Pressure sensors are often used to determine the interactions of fabric under varying conditions of pressure. By giving real-time data, these sensors can be used to determine the areas under critical stress concentrations and help in reducing drag. This also helps in improving the lifecycle of a product and increasing the service hours of a product, especially fabrics in this case. Temperature sensors are used to measure temperature changes and provide feedback on real-time interfacing between the fabric and changes in temperature. These sensors are generally extremely sensitive, providing excellent feedback for a little change in conditions. These sensors also help in producing materials that interact with the environment in various ways with varying temperatures [5]. These sensors provide data on heat distribution along the surface and give feedback on stiffness under cold conditions and deterioration under extreme temperatures.

Flow sensors help monitor real-time flow rates of water upon interaction with fabrics under different flow conditions. These are responsible for monitoring friction, otherwise drag, and ensuring the product is rolled out with minimal fabric drag and higher efficiencies. Different heights and thicknesses of ridges, along with distances between each ridge, can be analyzed using different flow conditions to identify the optimal condition for better usage. Smart sensors, depending upon the interactions with the surface, change the surface characteristics, and using computational properties, make changes in the surfaces to produce an optimal surface required for a particular condition. Along with interactions from the environment, these sensors can also be commanded from a computer, forcing a change in surface characteristics and analyzing it under various extreme conditions, providing real-time data with accurate results. Sensors and smart sensors together can provide us with real-time data under dynamic conditions, which helps manufacturers provide fabrics with optimal properties helpful in making human lives easier to live under various circumstances. Not only are these sensors used in sharkskin fabric applications, but they can also be used in various applications whenever real-time monitoring is required.

7.6 Actuators Used in Sharkskin Fabrics

Actuators used in sharkskin fabrics help in changing the structure of patterns according to the feedback provided to them by the sensors under dynamic conditions, such as changes in environmental conditions like pressures and temperatures. Piezo-electric actuators help in changing surfaces by providing a mechanical output for a change in electrical input [4]. Under various flow conditions, these help in making the dendrites present on the fabric surface protrude or flatten under different conditions so as to increase efficiencies by reducing the drag or by providing the grip required under extreme flow rates. All this data is collected in real time and later used in improving the surface characteristics [5, 11]. Shape-memory alloys are substances that change surface characteristics when temperature variations occur. These materials can react to changes in temperatures independently without any external inputs, making them an essential component when having a differential temperature condition.

Electroactive polymers change the shape and size of patterns, riblets, or dendrites whenever necessary under differential electric fields. These temporary changes help in maintaining high hydrophobicity, excellent reduction in drag, and improve overall efficiency and therefore human comfort. Micro-electro-mechanical system actuators are used to manage individual denticles. These are less explored due to their extremely small sizes, but they can provide rapid and individual responses and change the microstructural surface of fabric under varying environmental conditions. These devices consume very low power and allow precise control of the fabric's characteristics. Pneumatic actuators, when they encounter a fluid or air, help in shape changes. These are also called soft actuators and often help in orientational changes of dendrites, helping improve efficiencies under varying conditions. Sensors, along with these actuators, help maintain optimal results in real-time under varying conditions, helping in improving performance characteristics and improving efficiencies, along with hydrophobic properties. These devices produce independent responses without the need for external corrections, ensuring smoother and better experiences for users.

7.7 Conclusions

This study provides a theoretical overview of the structure of shark skin and its properties under various conditions. It also provides readers with the latest technological applications in making these structures, along with a brief overview of the types of sensors and actuators that can be used to create these surfaces artificially. Sharks have tooth-like structures on their skin called dendrites, which can naturally increase and decrease in size according to the conditions the shark swims in. Various shark species have different structures and shapes in their skin according to their environmental conditions. From a materials science perspective, shark skin combines properties such as rigidity, strength, and toughness with hydrodynamic properties. While denticle density does not have much significance on ultimate tensile strength, higher densities increase stiffness and reduce toughness. These findings highlight both the hydrodynamic and structural properties of shark skin. In drag-reducing applications, shark-inspired riblet structures have shown better abilities than many other recent innovations. Blade-type riblets, for instance, achieve drag reductions of up to 9.9%, significantly enhancing fuel efficiency in aviation and resistance reduction in pipe flow engineering.

Three-dimensional printing has redefined the replication of shark skin structures. This technology enables precise control over denticle size, morphology, and arrangement, creating opportunities for tailored applications. However, certain challenges in replicating the exact morphology of biological shark skin need some further improvement in technology. The hydrodynamic advantages of shark skin enhance not only drag reduction but also swimming performance and energy efficiency. Studies have shown up to a 6.6% increase in the speed of the shark and a 5.9% reduction in energy consumption, emphasizing its potential in aquatic and aerial applications. In conclusion, the research on shark skin patterns and its bio-inspired designs establishes a relation between biology, engineering, and material science, highlighting how nature's evolution can make complex hydrodynamic structures. The usage of these principles in human-made technologies has already shown great potential, and ongoing research promises even greater advancements.

7.8 Future Research Opportunities

Although the current research on shark skin has significantly increased our understanding of its unique hydrodynamic properties, several other challenges remain today. While biomimetic applications of shark skin have made considerable innovations, replicating its full-scale denticle structure remains a challenge, highlighting the need for continued research. The reliance on samples from deceased sharks provides valuable information but excludes the workings of living systems, such as muscle tension and stiffness, which likely play a critical role in the functional dynamics of shark skin during swimming. Future studies focusing on these characteristics could offer a more comprehensive understanding of these systems.

A much broader range of shark species could provide valuable insights into how variations in denticle pattern, size, and arrangement influence drag reduction and swimming dynamics. The bristling effect observed in the shortfin mako shark represents another fascinating area for investigation. While efforts have been made to mimic this phenomenon, full replication into biomimetic designs remains unclear. Advancements in 3D printing technology can give a more precise replication of shark skin. With innovations in multi-material 3D printing, it is feasible to create biomimetic skins that accurately mimic the morphology and mechanical properties of shark denticles. These capabilities, combined with improvements in computational modelling through advanced CFD tools and deep machine learning, can accelerate the development and testing of biomimetic designs.

By addressing the remaining research gaps and using several cutting-edge tools, future investigations can refine our understanding of shark skin's hydrodynamic efficiency, making innovative solutions inspired by nature's evolution, leading to sustainable advancements in technology and materials science.

References

1. Baeten, S.R.R., Kochovski, A., Jovanova, J., Sakes, A., Characterization of shark skin properties and biomimetic replication. *Bioinspir. Biomim.*, 19, 051002, 2024.
2. Ghimire, A., Dahl, R.B., Shen, S., Chen, P., Shark Skin Denticles: From Morphological Diversity to Multi-functional Adaptations and Applications. *Adv. Funct. Mater.*, 34, 2307121, 2024.
3. Gabler-Smith, M.K. and Lauder, G.V., Ridges and riblets: Shark skin surfaces versus biomimetic models. *Front. Mar. Sci.*, 9, 975062, 2022.
4. Fan, S., Han, X., Tang, Y., Wang, Y., Kong, X., Shark Skin—An Inspiration for the Development of a Novel and Simple Biomimetic Turbulent Drag Reduction Topology. *Sustainability*, 14, 16662, 2022.
5. Chen, D., Liu, Y., Chen, H., Zhang, D., Bio-inspired drag reduction surface from sharkskin. *Biosurf. Biotribol.*, 4, 39–45, 2018.

6. Domel, A.G., Saadat, M., Weaver, J.C., Haj-Hariri, H., Bertoldi, K., Lauder, G.V., Shark skin-inspired designs that improve aerodynamic performance. *J. R. Soc. Interface*, 15, 20170828, 2018.
7. Fu, Y.F., Yuan, C.Q., Bai, X.Q., Marine drag reduction of shark skin inspired riblet surfaces. *Biosurf. Biotribol.*, 3, 11–24, 2017.
8. Luo, Y., Liu, Y., Anderson, J., Li, X., Li, Y., Improvement of water-repellent and hydrodynamic drag reduction properties on bio-inspired surface and exploring sharkskin effect mechanism. *Appl. Phys. A*, 120, 369–377, 2015.
9. Chen, H., Zhang, X., Che, D., Zhang, D., Li, X., Li, Y., Synthetic Effect of Vivid Shark Skin and Polymer Additive on Drag Reduction Reinforcement. *Adv. Mech. Eng.*, 6, 425701, 2014.
10. Oeffner, J. and Lauder, G.V., The hydrodynamic function of shark skin and two biomimetic applications. *J. Exp. Biol.*, 215, 785–795, 2012.
11. Bixler, G.D. and Bhushan, B., Shark skin inspired low-drag microstructured surfaces in closed channel flow. *J. Colloid Interface Sci.*, 393, 384–396, 2012.

8

Spider Silk-Inspired Ropes and Fibers: From Nature's Design to Artificial Production

Abstract
Spider silk is one of the well-discussed natural fibers in the present-day world. Its exceptional physical and chemical properties make them useful in various fields. This chapter discusses the various properties of spider silk and their potential applications, especially as a rope with high toughness that can withstand fractures. The structure of natural spider silk is analyzed, and the possible artificial substitutes are also listed. Based on the raw material from which the artificial fiber is made, the fibers have different properties. There are different types of spinning mechanisms that can be used to produce artificial spider silk fiber from various raw materials. The methodology, advantages, and disadvantages of each fiber production process. Apart from the structural peculiarity, spider silk exhibits many other properties that can be used for varied purposes. Some physical properties of spider-silk are also discussed in this chapter. Finally, the potential fields where the spider silk can be used in the future have been identified.

Keywords: Spider-silk, spidroin, natural fiber, toughness, dry spinning, wet spinning

8.1 Introduction

Spider silk is a natural protein fiber with extraordinary properties. It has extraordinary properties in terms of optical, thermal, magnetic, water response, biocompatibility, toughness, strength, and so on. Among them, some of the very significant properties are toughness, strength, and elongation. Strength is the maximum stress a material can tolerate without failure and permanent

Jagadeesha T. and Sandip Kunar. Bio-Inspired Manufacturing Systems, (185–208) © 2026 Scrivener Publishing LLC

deformation [1]. Elongation is the ability of the body to deform plastically without failure. Toughness is the ability of a material to absorb energy before fracture occurs. These are some of the properties that are given importance while selecting engineering materials. Materials with these properties are used in many critical applications, like body armor, safety ropes, pressure vessels, artificial muscles, and the manufacturing of objects that may have to withstand impact loads. Spider silks are even better than steel, which is one of the most commonly used engineering materials; i.e., spider silk has very high toughness, so it will not fracture easily. Toughness is a combination of the strength and ductility of a material. Usually, the materials have either good strength or ductility. But spider fiber is one of the biofibers that shows good strength and ductility simultaneously. These properties of spider silk can be explained using micro and nanostructures [2].

Spider silk is the most elastic protein fiber in nature. It is also a very tough material with a 1.6 GPa of strength and can elongate up to 50% of its length. Despite these excellent mechanical properties exhibited by spider silk, it is not widely used in the current industries. The main reason for this is the difficulty of producing it. Usually, silk from silkworms and other animals is extracted after breeding the silkworm and harvesting the silk fiber. But it is possible to breed spiders like silkworms because spiders are territorial creatures; i.e., there is a high possibility that they may kill each other. So, the method is to produce it artificially. This leads to a shortage of spider silk. Now, there are different spider silk-inspired materials produced from proteins, non-protein polymers, and composites [3, 4].

Spiders have an efficient spinning mechanism within them. During spinning, the spinning fluid undergoes both physical and chemical transformations. The pH inside the gland and the shape of the spinning tube help to achieve this. Finally, the highly concentrated spun protein comes out from the legs of the spider. It gets solidified in the air and forms spider silk. Some among them are dry spinning, wet spinning, electrospinning, and microfluidic spinning. Other methods, like 3D printing and film molding, are also used to produce tough fibers. Even though these processes have been developed, they have serious disadvantages in terms of efficiency, cost, and quality [5].

8.2 Primary and Secondary Structures of Spider Silk

The uniqueness of spider silk lies in its fundamental structure. A deeper understanding of its complicated structure is essential to realize how it has far better mechanical properties than any synthetic fibers. Spider silk has a multi-stage structure: a primary skin core structure [6] and a secondary nanofibril structure. The protein that makes spider silk is called a spidroin. When spider silk is observed, a series of nanofibrillar structures, which are highly twisted and interconnected, can be seen. This interconnection plays a key role in defining silk's high toughness. Further enlargement of fibrils reveals that each of these small fibrils contains anti-parallel β sheet crystals and α helix in random coils. Every protein is nothing but a specialized sequence of amino acids. Similarly, our spidroin is also a highly repetitive combination of a simple amino acid sequence. The fundamental repeating unit of spidroin is A-GGX, where A stands for Alanine, G for glycine, and X can be tyrosine, leucine [7].

This repeating unit can be split into two: a poly-alanine block and glycine-rich domains. The poly-alanine block typically consists of four to seven alanines, which form the antiparallel β sheets in the crystalline region. And in amorphous regions, glycine-enriched GGX randomly repeats in the form of a coil. These crystalline β sheets and amorphous α helix will arrange on the long axis of filament fibers to form the fibrils. Multiple H-bonds ensure the connection between the two regions. Studies reveal that the crystalline region has comparatively better orientation, and it is the main reason for incomparable strength and higher toughness. An amorphous region is responsible for its elasticity and ductility. So, spider silk is generally considered a semi-crystalline fiber in which the crystallites are dispersed in an amorphous matrix [8]. The high orientation of the crystalline β sheets is qualitatively determined by methods like Raman spectroscopy.

8.3 Orb-Weaver Spider and Dragline Silk

As of September 2024, around 52000 species of spiders are found in the world, out of which 8000 are orb weavers. Due to the unique features of the silk from these spiders, worldwide, several scientists are researching different orb weavers, like Paul L. Babb, who has been studying the dragline silk of Darwin's spider (*aerostris darwini*), whose dragline is one of the toughest known biomaterials on earth [9]. A typical female orb weaver has multiple silk glands. As of now, seven different kinds of anatomically different silk glands have been identified. Each of these glands makes a different kind of silk, and their use is also different.

a) Major ampullate (MA) gland – silk from the MA gland is used to make radii and structurally strong parts of their web. Silk from the major ampullate gland is called MA silk or dragline silk. The same silk is used by spiders as a safety silk to keep them safe on walls. The strongest silk is MA silk.
b) Minor ampullate gland – it is the inelastic, temporary line used by spiders as they crawl around their orb during its making.
c) Piriform gland – it is that special kind of silk that is used to cement the other silks together; it acts as a knot at each silk junction.

d) Flagelliform gland – Flagelliform silk is the stretchiest silk of all. It acts as the capture spiral of the web. This is the silk that withstands the impact of flying insects or prey. But it is not sticky itself.
e) Aggregate gland – This sticky, viscous silk makes the flagelliform silk sticky. It acts as a drop of glue that goes along the flagelliform silk.
f) Aciniform gland – it is the silk that spiders make quickly to wrap up the prey. It is also the soft inner material of their egg cases.
g) Tubuliform gland – To protect their eggs in winter, they make strong outer silk from this gland.

These are the seven major silk glands of orb weaver spiders. There may be other types of silk glands that are still unidentified. But, among these, the most researched and most suitable silk is dragline silk/MA silk for artificial silk-inspired ropes.

8.4 Major Ampullate Silk

Among the seven different types of spider silks, dragline silk has the most suitable mechanical properties. This silk, produced by the MA gland, has significantly higher tensile strength and higher elasticity. In a comparison study between equal-diameter steel wire and dragline silk, it was found that dragline silk is twice as strong as steel, having an almost 480 MPa breaking stress. Dragline silks are easy to obtain; they exhibit a unique combination of elasticity and high-tensile strength [10].

Studies revealed that Caerostris Darwini's MA silk strength and toughness are 1651MPa and 354 MJ/m^3 [1, 6]. Among the seven silks, dragline silk and flagelliform silk absorb more energy. They also exhibit supercontraction when exposed to water. When exposed to water, they contract up to 50% of their original length. This ability helps them to increase tightness under humid conditions [1, 7].

Dragline silk mainly consists of two types of spidroin proteins. Major ampullate spidroin-1 (MaSp-1) and Major ampullate spidroin-2 (MaSp-2). Both have the same crystalline polyalanine structure, but different amorphous parts. In MaSp-1, the amorphous part mainly consists of glycine-rich tripeptides (GGA, GXG). In MaSp-2, the amorphous part is dominated by proline-rich pentapeptides. Both sequences are presented in Table 8.1 and Table 8.2, respectively.

Considering the entire chain of amino acids, typical dragline silk mainly consists of glycine at 37.24%, alanine at 17.6%, proline at 15.7%, and glutamine at 11.49%. Dragline spidroin protein mainly consists of highly repetitive core domains and non-repetitive NTDs (N-Terminal Domains) and CTDs (C-Terminal Domains). Each repeating core domain is flanked on both sides by NTDs and CTDs. The repetitive core domain gives strength to spider silk by forming crystalline β sheets and NTD's and CTD's contribute to fiber formation [11]. During the web-spinning, it was found out that a pH drop to 6.3 along the spider's spinning apparatus, and altered salt composition and shear force are responsible for the conversion of silk fluid to solid silk. From the research, it is found out that the N-terminal

Table 8.1 Amino acid sequence of MaSp-1.

Nephila clavipes
---GPG--QQGPGGYGPG-QQGPGGYGPGQQGPSGPGSAAAAAAAA
Nephila madagascariensis
GPG-QOGPGGYGPGQQGPGGYGPGQQGPSGPGSAAAAAAA-
Nephila senegalensis
---GPG--OOGPGXY--GPSGPGSAAAAA—

Table 8.2 Amino acid sequence of MaSp-2.

Nephila clavipes
GGA--GOGGYGGLGSQGA-GRGGLGGQ-GA-GAAAAAA
Nephila madagascariensis
GGA--GOGGYGGLGSQGA-GRGGYGGQ-GA--GAAAAAA
Nephila madagascariensis
GGA--GQGGYGGLGGQGA

domain is responsible for the pH sensitivity and the C-terminal domain contributes to silk formation [10].

8.5 Nanofibrillar Structure

The nanofibrillar secondary structure of dragline silk also plays a key role in giving the silk its mechanical properties. Atomic force microscopy revealed that dragline silk had twisted bundles of nanofibrils. Studies also show that its diameter is 35 nm. The highly twisted nanofibrils are also interconnected. This interconnection leads to high toughness and strength. Another advantage of the nanofibrillar arrangement under stretching, impact is significantly reduced due to the quickly formed nano-cracks. If there are no nanofibril structures, then large cracks might have formed. There are several experimental verifications of this nanofibrillar structure, giving the strength for dragline silk.

To understand how it can withstand higher stress and higher load, it is important to observe more of the β-sheet arrangements within it. As described in the structural analysis, antiparallel β-sheet crystals at the nanoscale contain poly (Gly-Ala) and poly-Ala repeats. The β-sheet crystals bonded by hydrogen bonds, roughly constitute 10–15 % of silk. But for decreased ordered structure, they can fill up to 50% of the silk volume. This less orderly structure helps silk protect itself during stretching. During stretching, β-sheet crystals reinforce the partially ordered molecular chain by forming interlocking regions, which transfer load during lateral loads [12]. So, the β-sheet crystals provide cohesion between those strands, which helps the amorphous domain to stretch largely. In this way, β-sheet crystals provide improved resilience, stiffness, and even better fracture toughness. By computational experiments regarding bending and pulling out, it is found that β-sheet crystals that have a size lesser than a critical size have significantly higher stiffness and toughness. It is because they are under uniform shear, and cooperative rupturing of hydrogen bonds will occur. So, larger nanocrystals will be more prone to breaking when compared to smaller nanocrystals. It is found that nanocrystal dimensions below 3 nm increased the ultimate stress by multiple folds.

8.6 Skin Core Structure of Dragline Silk

Through an atomic force study of spider silk, it is found that dragline silk exhibits a skin-core structure. They did the atomic force study of MA silk and found out its chemical and morphological composition. It is found that from exterior to interior, there are a total of five clearly visible layers. On the exterior side, there is a lipid layer, a glycoprotein layer, and a skin layer. And in the interior core side, there are outer core layers and inner core layers.

Among these skin layers, MaSp-1 and MaSp-2 are only present in two core layers; also, their distribution is heterogeneous [7]. It was found that MaSp-1 is uniformly distributed across the outer core layer, and MaSp-2 is missing in the periphery but tightly packed in the inner core, which has a diameter of 3–3.5 micrometers. This shows that in the outer core, MaSp-1 is enriched, which in turn increases the β-sheet content. So, it contributes to more strength and toughness. Whereas, in the inner core, the presence of tightly packed MaSp-2 loosens the packed matrix and provides higher elasticity.

8.7 Factors Affecting the Mechanical Properties of Spider Silk

8.7.1 Reeling Speed

Research revealed that the increased spinning speed of a spider is directly correlated with increased breaking stress and Young's modulus. This decrease in elasticity is because, at high spinning speeds, molecules of liquid crystals become more oriented. This increased orientation will cause silk to be stiffer and stronger, leading to less elasticity. So, at a higher spinning speed, β-sheet crystals will be highly oriented, and protein macromolecules in an amorphous state are highly aligned, producing high stiffness [13].

When the spinning speed is increased, spiders produce silk with high intra-molecular β-sheet crystals in silk fibrils. With variable spinning speed, spiders can regulate the amount of β-sheet crystals in their silk and, in turn, adjust their silk stiffness. In fact, spiders have muscular control over their spinnerets [14]. As a resistive mechanism, they also use their legs, which in turn control the speed of spinning. A fast-spinning speed leads to stiffer spider silk, which can hold the weight of a spider during vertical jumping or predator attack.

8.7.2 Heat and pH

Mechanical properties of silk and heat are also correlated. Heat significantly helps silk align its crystals while spinning. Spiders maintain an optimal pH level of 6.3 to increase the mechanical properties of silk.

8.7.3 Shear Force During Spinning

During the passage through the spinning duct, silk fluid experiences significant shear forces due to the small diameter of the ducts. This shear stress will lead to increased aggregation of granular proteins. The shear force also affects the mechanical properties of spider silk by determining the molecular orientations of silk proteins just before they are pulled out.

8.7.4 Some Environmental Factors

8.7.4.1 Wind

In stronger wind conditions, spider webs should have high strength and high extension to defend themselves from breaking. For this, the web must have high-strength dragline silk/MA silk. The amino acid composition of high wind area spiders, like G *C. ginnaga*, contains more glycine and a lower glutamine percentage, which in turn leads to high tension and breaking energy. In a detailed study of the wind's effect on spider silk [15], it was found that the wind changed the number of β-sheets and their arrangement than the amino acid composition. The size and density of β-sheets mainly control the tensile strength and toughness of silk. A larger β-sheet will lead to decreased toughness of silk. According to the current study, wind can also affect the frictional stress at the spigot valve and shear force during the pulling out.

8.7.4.2 Temperature

The body temperature of ectothermic animals and their movement speed are highly correlated. It was found that dragline silk has higher toughness in a cooler environment compared to a hotter ambient temperature. Due to the presence of intermolecular H-bonds, MA silks show significant heat resistance up to 371 degrees Celsius. Dragline silk of spiders, like *N. edulis*, shows higher toughness even at very low temperatures (-60 to 0 degrees Celsius).

8.7.4.3 Humidity

The phenomenon of super contraction, as explained earlier, is that MA silk contracts up to 50% of its original length; this contraction helps the silk generate an excess of 50 MPa ultimate strength. H-bonding is responsible for the alignment of crystals in the amorphous region along the axis of the MA silk spider. Water can disrupt the spidroin H-bonding and allow the spidroin to attain a more disoriented state. This leads to cyclic swelling and contraction in high-humidity conditions. Relative humidity also affects the glycoprotein capture and elastic mobility of dragline silk.

8.7.4.4 Exposure to UV Radiation

From the study, it is found that prolonged exposure to UV rays causes the degradation of spider silk [7]. It may be the reason for spiders renewing their webs daily due to continuous exposure to sunlight. Certain conditions, such as acid rain and acid mist, also degrade spider silk.

8.8 Raw Materials

8.8.1 Recombinant Spidroins

It is one of the most common raw materials used for the manufacturing of artificial spider silk. As mentioned earlier, the territorial behavior of spiders hinders the chance of producing spider webs by breeding spiders. Spidroin is the main protein present in spider silk. Recombinant spider silk is produced from hosts, like insect cells and the mammary glands of goats, by modern genetic engineering, like gene designing, cloning, and protein expression. Later, this obtained recombinant spidroin is passed through a spinning process to produce the silk fiber. Using this, researchers could produce spider silk that is comparable to natural spider silk. It was also found that the molecular weight of spidroins is a very important factor in determining the mechanical properties of the silk produced. High-molecular-weight spidroins can bring up enhanced mechanical properties. Table 8.3 consolidates the tensile strength, extensibility, and toughness of artificial spider silks obtained from different types of recombinant spidroin.

Table 8.3 Properties of spider silk produced by recombinant spidroin through different methods.

Method	Tensile strength	Extensibility	Toughness
Wet spinning of recombinant spidroin with 6000% post-stretching	370 ± 59 MPa	110 ± 25%	183 ± 33 MJ/m^3
Short major ampullate spidroin	282 ± 66 MPa	-	144 ± 44 MJ/m^3
Heterodimerized spidroin	834 ± 34 MPa	32 ± 1%	143 ± 6 MJ/m^3
High-molecular-weight spidroin	508 MPa	15%	-

8.8.2 Non-Spidroin Proteins and Polypeptides

The microstructure of spider silk and silkworm silk has many similarities. Both have anti-parallel β-sheet nanocrystals and an amorphous matrix consisting of α-helices and random coils. The combination of these two elements produces a synergistic effect to form crosslinks, which is one of the important reasons for high strength and toughness. This idea can be extended to non-spidroin proteins. So, regenerated silk fibroin (RSF) can be used for the manufacturing of spider silk. Similarly, many methods are explored by researchers to produce artificial spider silk from non-spidroin proteins.

Another important method for producing spider silk is by using polypeptide chains. Spidroin is a natural protein whose structure is known to us. All proteins are made up of polypeptides. So, we can mimic a protein just by imitating the same polypeptide chain structure. The most common method for this polypeptide chain consists of two steps. A chemoenzymatic polymerization followed by post-polycondensation using phosphoric acid. In the first step, two polypeptides are produced. One is polyalanine and the other is polyglycine. Among them, the primer forms a β-sheet and the latter

Table 8.4 Properties of spider silk produced from non-spidroin proteins and polypeptides using different methods.

Method	Tensile strength	Extensibility	Toughness
Regenerated silk fibroin fiber by microfluidic spinning	614 MPa	27%	-
Collagen fiber by microfluidic spinning	383 ± 85 MPa	-	-
Regenerated silk fibroin/ TiO2 composite fiber	218.5 MPa	88.3%	93.1 ∓ 27.1 MJ/m3
Dry spinning of RSF and graphene oxide	435.5 ± 71.6 MPa	35.9%	-
Spinning of Pseudo Protein polymer fibers	100 MPa	-	387 MJ/m3

forms the soft amorphous region. These two polypeptides are ligated in the second stage, which results in a multiblock peptide. Table 8.4 shows the strength, extensibility, and toughness of artificial spider silk produced using different methods.

8.8.3 Polymer Materials

Various fibers and bulks can be made by the physical or chemical crosslinking of hydrogels in water. Hydrogel fibers, which are similar to the core-shell structure of natural silk, can be used to produce artificial spider silk. The silks produced thus can be reinforced using methods like ion doping and twist insertion. Spinning hydrogel fibers is not easy due to their 3D structure. So, the hydrogels are drawn using a metal wire dipped in a hydrogel bulk. The diameter of the wire can be precisely controlled using this method by altering the dipping depth.

Apart from hydrogels, polyurethane fibers can also be used to produce spider silk. Bacterial cellulose nanofibers also exhibit a high degree of polymerization and crystallinity, which makes it a good choice for making fibers. They are obtained by bacterial fermentation.

8.9 Methods of Fiber Production

8.9.1 Dry Spinning

Dry spinning is a simple method. In this method, the fiber is produced by directly extruding the fiber into the air and allowing it to solidify. A polymer solution containing polymer and solvents is taken in a storage unit. Then it is passed through a filter and then through a metering pump or a doping pump to a spinneret. A spinneret is a plate with several holes through which the solution is forced. Metering pumps can move a very accurate amount of liquid at a given time. Its flow rate can also be adjusted. The solution coming out of the spinneret is passed through an evaporation chamber. A heating jacket is provided as a lining in the evaporation chamber to provide heat. The high-density cool air enters the chamber from the bottom, gets

Table 8.5 Advantages and disadvantages of dry spinning.

Advantages	Disadvantages
Continuous production	Flammable solvents are used
High spinning speed	Requires solvent recovery
Large yield	Fiber cross-sections may vary slightly
Simple post-spinning process	Large-scale spinning is difficult

heated, and leaves through the outlet at the upper portion. Along with the air, the volatile solvent from the fiber evaporates. By this time, the filament is produced. Now, the produced filament is passed through certain godets and rolled up to collect the fiber spool. The filament is passed through the godet to ensure that the filament is evenly twisted and aligned. Major advantages and disadvantages of dry spinning are given in Table 8.5.

8.9.2 Wet Spinning

This is the most used method to produce artificial spider silk. The chemical properties and dyeing properties of wet spinning fibers are inferior to those of dry spinning fibers. Here, also, the process is almost the same as dry spinning, but the polymer solution is injected into a liquid bath from the spinneret instead of the evaporation chamber. In this process, the diffusion takes place twice. First, the diffusion of solvent in the polymer solution, then the diffusion in the liquid bath. During this process, the polymer will aggregate and precipitate from the stock solution. The liquid bath is called a coagulation bath because precipitation happens to a polymer when passing through it. Later, this is solidified to get the primary fiber. After that, the fiber is passed through godets and collected as a spool, as in dry spinning. The post-stretching process in wet spinning helps promote β-fold conformational transformation and enhances molecular chain orientations, which improves the mechanical properties of the fibers. Advantages and disadvantages of wet spinning are given in Table 8.6.

Table 8.6 Advantages and disadvantages of wet spinning.

Advantages	Disadvantages
Biocompatibility	The spinning speed is low
Environment friendly	Should be washed to remove impurities
Can adjust fiber properties	Complicated process
Easy to do compared to other processes	High cost

8.9.3 Electrospinning

There are two methods of electrospinning: one is with a needle, and the other is without a needle. This process starts with the polarization of the polymer solution. A force from the electric field is used to pull down the polarized solution to spin, and is deformed by the high-voltage electrostatic field. Initially, the polymer solution forms a droplet at the tip of the needle/nozzle. Later, when the electrostatic force becomes greater than the surface tension of the droplet formed at the tip, the solution is sprayed at very high speed as a jet. Volatile solvents in the solution evaporate within this small distance, and the jet solidifies and gets collected in a collector.

Electrospinning without a needle is somewhat different from the above process. In the process without a needle, the electrode is placed inside the spinning fluid and is rotated in an electric field of a certain voltage to get the spinneret. In this method, we can spin a bulk amount of fiber in a shorter amount of time. However, since the solvent is directly exposed to the air, it evaporates. Gradually, the solvent concentration reduces to a level beyond the threshold level, which is required to produce the silk. There are different kinds of rotating spinnerets used for needleless spinning. Some of them are: Rotating drum spinneret, rotating ball spinneret, rotating disk spinneret, rotating coil spinneret, rotating gear spinneret, and rotating wire electrode spinneret.

In these methods, we can have complete control over the properties of the fibers obtained by changing the needle type and technological parameters, such as voltage and distance. One can customize the thickness, surface area, diameter, and many other properties. Nanofiber membranes can be produced using this method. But with all these advantages, the low efficiency of this process makes it difficult to be mimicked on an industrial scale [3].

8.10 Microfluidic Spinning

Microfluidics is used in microfluidic spinning to achieve spinning. The spinning is made possible by controlling the formation and flow of droplets. This method is better than the widely used wet spinning because of its controllability, selectivity, and repeatability. Polymer solution and gel solution are required in this process. A microfluidic chip is used in this method to force these two solutions into two adjacent microchannels and form an interface. During the flow, the mixing of both solutions is controlled. As a result, polymer molecules in the polymer solution form a fiber structure. After that, the fiber thus produced will be fixed on the gel solution by its gel action. Thus, the fibers of the required size are made. this method can be used for the preparation of different types of microfibers, such as solid fibers, core-shell structures, hollow structures, and multi-component structures.

8.10.1 Novel Spinning Methods

Apart from the above-mentioned traditional methods, new methods have been developed by researchers to produce artificial spider silk. Some examples include melt spinning, dry-jet wet-spinning, liquid crystal spinning, reaction spinning, centrifugal spinning, air gap spinning, and instantaneous release pressure spinning.

8.11 Physical Properties of Spider Silk and Inspired Artificial Fibers

Rather than these incomparable mechanical properties, spider silk also exhibits some other unique physical properties, too. Water response, light transmission, torsional energy dissipation, and thermal conductance are some of them.

8.11.1 Water Response

Due to its high adsorbing capacity, there is a high correlation between silk and water. There are water-repelling peptides coating on spidroins. The behavior of spider silk changes significantly in a moist environment. Not just supercontraction, the hydrophilicity of spider silk also helps it capture prey. Novel solid-liquid behavior of spider silk under contact with water [6], which helps the web to contract and, in turn, lift more weight. This sensitivity to humidity can be converted into an advanced humidity sensor. The humidity response of spider silk is also utilized in artificially inspired fibers. A research group made a highly sensitive humidity sensor, which is a composite of methyl viologen polymer and hydroxyethyl cellulose. The cross-linked H-bond between these two structures is disrupted in highly humid conditions, like in spider silk, which, in turn, acts as a humidity sensor.

8.11.2 Light Transmission

In daytime, spider silk showed comparatively higher transmittance. The unique biocompatibility, flexibility, and light transmissibility of spider silk make it an efficient and biocompatible optical fiber. Straight spider silk has an optical loss of 10.5 dB/cm [1]. This opens the possibility of incorporating spider silk-inspired fibers in photonic chips and all. So, spider silk is very promising for photonics applications, as well. The artificial fiber prepared by the Wang group using genetically engineered spidroin exhibits an optical loss of 0.9 dB/cm in air, and the RSF/CNF hybrid fiber, which mimics the spinning apparatus of a spider, exhibited an attenuation of 1dB/cm.

8.11.3 Torsional Energy Dissipation

During hazardous situations, the spider jumps from the walls with a dragline silk attached to its standing point. But it never swings about the initial point and never rotates to and forth when torsionally triggered. This is due to its high energy-damping capacity and impressive shape memory behavior. It has a comparable damping capacity to that of Kevlar, copper wire, and shape memory Ni-Ti alloy. The dragline silk has a high torsional damping capacity. This is due to a special arrangement of β-sheet crystals. These β-sheet crystals maintain the initial shape by acting as a temporary node. By mimicking the shape memory capacity of spider silk, shape memory fibers from recombinant MaSp-2 spidroin were made, which showed comparable results with spider silk. This fiber showed excellent shape flexibility.

8.11.4 Thermal Conductance

As mentioned in the structural analysis, there are better-aligned and oriented β-strands in the crystalline region. This, in turn, increases the toughness and thermal conductance of silk. The structure of spidroin, such as the conserved repetitive units, NTDs, and CTDs, also contributes to the high thermal conductivity of spider silk. Under strain, their orientation is significantly improved, which increases the thermal conductivity of silk. Research also shows that an increase in β-strands of spidroin also leads to higher thermal conductivity.

8.12 Applications of Spider Silk-Inspired Fibers

There are many engineering applications of the spider silk-inspired fibers. Their unique qualities and properties offer many advantages and applications. Some of them are explained in subsequent sections.

8.12.1 Construction and Civil Engineering

Earthquake-resistant material – with its unique combination of lightweight and highly strong nature, spider silk-inspired fiber is a perfect choice for flexible composite materials that can increase the structural integrity of buildings. High-performance cables – spider silk-inspired fibers are the perfect replacement option for traditional cables used conventionally like steel cables. It can be used as cables for elevators and suspension bridges. Their high strength-to-weight ratio makes them ideal for cables. Lightweight reinforcements – for reinforcing materials like glass, concrete, spider silk-inspired fibers can be used. It can significantly improve strength without adding extra weight.

8.12.2 Electronics and Energy Transmission

Optical fibers – their high capacity for light transmission and less attenuation make them a perfect candidate for optical fibers for high-speed data transmission. Its lightweight nature is an added advantage. Thermal management systems – spider silks are highly conductive in nature. This high conductivity makes them suitable for improved heat dissipation systems in machines. So, they can act as perfect cooling systems. Electromagnetic shielding – on coating with conductive materials, spider silk can act as an electromagnetic shielding device, which is also lightweight. This will help to shield the devices from EM interference.

8.12.3 Robotics

Artificial muscles – their unique combination of strength and elasticity is a major requirement for muscles. So, spider silk-inspired fibers can act as actuators for robots, which can do the jobs of muscles more efficiently. Robotic exoskeletons – again, the strength and elasticity properties of spider silk combined can significantly improve the exoskeleton's design. It can increase mobility and reduce fatigue in users. High-performance grippers – the incomparable elasticity and torsional energy dissipation capacity of spider silk make it a good choice for adaptive grippers in the robotic industry.

8.12.4 Aerospace

Weight reduction in spacecraft components – spider silk can replace several commonly used heavier metals in spacecraft due to its high strength and toughness, which in turn reduces weight and improves fuel efficiency. Deployable structures – they can be used as deployable rods for solar sails or antennae. Its high extreme temperature withstanding capacity makes it possible. Space suits – their high flexibility, durability, and toughness make them the most suitable material for protective space suits in the harsh environment of outer space.

8.12.5 Agriculture

Crop protection nets – their super strength, elasticity, and lightweight nature can be utilized to make a protection net for plants. This will protect plants from pests, birds, and even extreme weather. Seed coating – As a seed coater, it can protect seeds from extreme conditions and also provide resistance to diseases. Insect trapping – their super sticky nature can be utilized to make traps for unwanted insects, which improves the efficiency of farming.

8.12.6 Other Applications

Skincare products- They are elastic and lightweight, which shows the possibility of integrating them in skincare products for skin repair and moisturization. Their hydrophilic nature and super contraction phenomena can also be used here for skin-tightening. Shock-absorbing material – in defense applications, it can be used as a shock-absorbing material to protect highly sensitive equipment. Climbing ropes – the high tensile strength and elastic nature of spider silk make them a perfect candidate for unbreakable climbing ropes. Surgical threads – they are light in weight, can have significant strength for a smaller thickness fiber, and are biocompatible. These properties make them a perfect choice for surgical threads.

8.13 Conclusions and Future Scope

A typical spider silk is 100-1000 times thinner than human hair, still the mechanical properties it shows are highly impressive. So just imagine what all applications it has, if possible to make it in higher thickness artificially. Aim and progress made by current studies signify that spider silk is a very promising fiber that can shape the future of human life. The complex micro and nano-level structures enable the spider silk to have a stronger, better, toughness than steel, Kevlar, or any other synthetic fiber materials. It is also super stretchy. In fact, Spider silk is a natural material that outperforms almost any synthetic material in its combination of strength and elasticity. Not only the strength, toughness, and ductility, but also there are a variety of properties shown by the spider silk which are really amazing. Some of them are water response, magnetic property, optical property, biocompatibility, torsional energy dissipation, etc. When its toughness makes it possible to make impact-resistant objects with it, while its biocompatibility showcases its applications in the medical field.

Even though spider silk is a material with immense potential, it is not seen used on an industrial scale. The scarcity of natural spidroin and the difficulty of producing artificial spider silks are the major reasons for this. Researchers have explored various methods for the production of spider silks, of which wet spinning is the most used. However, these processes are costly and are difficult to replicate on an industrial level economically [15, 16].

Today researchers are looking for an economical process to produce artificial spider silk fiber which is economical and for which the raw materials will be readily available [17]. If the issues related to both of them, raw material and process, are sorted out, spider silk will be one of the most widely used fibers in various fields. Once the materials inspired by spider-silk are able to be produced in industrial scale, it will revolutionize the fiber industry.

References

1. Li, J., Li, S., Huang, J., *et al.*, Spider silk-inspired artificial fibers. *Adv. Sci.*, 9, 5, 2021. https://doi.org/10.1002/advs.202103965.
2. Khan, A.Q., Guo, W., Li, S., *et al.*, Innovations in spider silk-inspired artificial gel fibers: Methods, properties, strengthening, functionality, and challenges. *SusMat*, 4, 3, 2024. https://doi.org/10.1002/sus2.205.
3. Understanding material strength, ductility and toughness. https://efficientengineer.com/, last accessed 2024.
4. Su, Y., Shi, S., Wang, C., *et al.*, Spider silk-inspired tough materials: Multi-pathway synthesis, advanced processing, and functional applications. *Nano Today*, 55, 102188, 2024. https://doi.org/10.1016/j.nantod.2024.102188.
5. Creating novel fibres from synthetic spider silk, Advertisement feature.
6. Elettro, H., Neukirch, S., Vollrath, F., *et al.*, In-drop capillary spooling of spider capture thread inspires hybrid fibers with mixed solid–liquid mechanical properties. *Proc. Natl. Acad. Sci. U.S.A.*, 113, 22, 6143–6147, 2016. https://doi.org/10.1073/pnas.1602451113.
7. Kitagawa, M., Katsumi, S., Wakoh, Y., *et al.*, Structure and deformation of dragline and capture threads of orb-web in *Nephila clavata*. *J. Soc. Mater. Sci., Jpn.*, 46, 5, 470–474, 1997. https://doi.org/10.2472/jsms.46.470.
8. Blamires, S.J., Wu, C., Sheu, H., *et al.*, Uncovering spider silk nanocrystalline variations that facilitate wind-induced mechanical property changes. *Biomacromolecules*, 14, 10, 3484–3490, 2013. https://doi.org/10.1021/bm400803z.
9. Cohen, E. and Moussian, B., Extracellular composite matrices in arthropods, in: *Springer eBooks*, https://doi.org/10.1007/978-3-319-40740-1.
10. Askarieh, G., Hedhammar, M., Nordling, K., *et al.*, Self-assembly of spider silk proteins is controlled by a pH-sensitive relay. *Nature*, 465, 7295, 236–238, 2010. https://doi.org/10.1038/nature08962.
11. Lefèvre, T., Rousseau, M., Pézolet, M., Protein secondary structure and orientation in silk as revealed by Raman spectromicroscopy. *Biophys. J.*, 92, 8, 2885–2895, 2007. https://doi.org/10.1529/biophysj.106.100339.

12. Keten, S., Xu, Z., Ihle, B., *et al.*, Nanoconfinement controls stiffness, strength and mechanical toughness of β-sheet crystals in silk. *Nat. Mater.*, 9, 4, 359–367, 2010. https://doi.org/10.1038/nmat2704.
13. Shi, Y., Zhu, Y., Qian, Z., *et al.*, Artificial spider silk materials: From molecular design, mesoscopic assembly, to macroscopic performances. *Adv. Funct. Mater.*, 2024. https://doi.org/10.1002/adfm.202412793.
14. Babb, P.L., Gregorič, M., Lahens, N.F., *et al.*, Characterization of the genome and silk-gland transcriptomes of Darwin's bark spider (*Caerostris darwini*). *PLoS One*, 17, 6, e0268660, 2022. https://doi.org/10.1371/journal.pone.0268660.
15. Allmeling, C., Radtke, C., Vogt, P.M., Technical and biomedical uses of nature's strongest fiber: Spider silk, in: *Springer eBooks*, pp. 475–490, https://doi.org/10.1007/978-3-642-33989-9_36.
16. Ranade, V., Spider silk inspires a new route to organic magnets. *MRS Bull.*, 2024. https://doi.org/10.1557/s43577-024-00667-z.
17. Lewis, R.V., Spider silk: Ancient ideas for new biomaterials. *ChemInform*, 37, 48, 2006. https://doi.org/10.1002/chin.200648276.

9

Biomimetic Design and Fabrication of Hummingbird-Like Micro Unmanned Aerial Vehicles

Abstract

Hummingbirds, renowned for their agility and maneuverability, are among the few vertebrate bird species examined in this chapter. The focus is on their maneuvering capabilities and morphological characteristics, as referenced in various cited studies. Limited open-domain research on the morphology of hummingbird-like drones presents an opportunity to explore innovative unmanned aerial vehicle (UAV) designs inspired by these birds, offering valuable insights into the development of micro UAVs. This chapter examines key aspects of manufacturing, including types of UAVs and the challenges related to miniaturization. These factors provide insights into hummingbird-inspired drones, particularly regarding control mechanisms, stability, and operation in confined spaces. A major challenge is selecting a small drone capable of replicating specific wing movements. The integration of flexible body components to enhance maneuverability and adaptability is also explored. Applications of hummingbird-inspired drones across various sectors are discussed, highlighting their advantages while acknowledging limitations in design and functionality. Ethical and legal considerations are briefly included to support responsible development. No physical models were created; the study is based on a literature review and conceptual analysis of drone designs. This chapter presents a comprehensive overview of the design and manufacturing principles of miniature UAVs inspired by nature, emphasizing how biological forms and flight mechanisms can inform advanced engineering solutions.

Keywords: Hummingbird-inspired drones, biomimetics, micro UAV, flapping wing mechanism, miniaturized manufacturing

Jagadeesha T. and Sandip Kunar. Bio-Inspired Manufacturing Systems, (209–238) © 2026 Scrivener Publishing LLC

9.1 Introduction

Recent trends show that research in biomimetics is being done extensively, as it promises sustainable engineering solutions. There is a major difference between a biological system and a man-made system. Man-made systems always depend on materials, energy, methods, and sustainability. Biological systems, on the other hand, effectively use natural information, space, and hierarchical structures to optimize resources based on the restrictions imposed upon them. Natural systems are unable to adjust themselves as humans do modifications. Comparing the costs and efficiency of drones, flying is the favorable method in terms of movement over terrestrial landscapes. Moreover, the unique maneuverability during flying, hovering, and distinct wing dynamics are other factors that make this topic particularly interesting to us. This chapter focuses on the unique and special behavior in the maneuverability of hummingbirds. When they exhibit such behavior together with exceptional control over their movements, such as hovering and rapid directional changes, making them ideal models for bio-inspired drones. The aim of this is to get a comprehensive understanding that will bridge the gap between bio-inspired technology and engineering solutions [1].

A significant aspect of this study involved evaluating the feasibility of designing and manufacturing a hummingbird-inspired drone. Material and fabrication are also important. To construct a lightweight, durable, and flexible drone. One of the critical challenges of concern was miniaturization. Furthermore, to ensure stability and precise control, advanced position and stability control systems are adapted for these concepts of small-scale drone applications. Hummingbirds move their wings in a specific manner during their hover and maneuver situations. The analysis of reconstruction methods and computational fluid dynamics (CFD) simulations is important to study the aerodynamics of bio-inspired drones [2]. Forces like aerodynamic forces and wing movements were crucial for getting accurate and proper maneuvers. These techniques enabled a detailed analysis. When a drone operates in a place like a forest or at a high altitude, there is a severe risk of bird attacks due to the drone's resemblance to a real bird. This consideration is a major issue where other kinds of protection are not possible. By addressing these challenges, the study aims to advance the development of biologically inspired drones that can perform complex maneuvers while overcoming real-world limitations.

9.2 Importance of Hummingbird-Inspired Drones

Designing for small drones is possible, but manufacturing drones with smaller sizes is always challenging, and no one can guarantee success. A necessary characteristic of a bio-inspired micro-aerial vehicle (MAV) is its ability to perform quick and safe yaw turn movements. The effectiveness of such MAVs becomes crucial when they operate in confined spaces and several obstacles are present [3]. Hummingbirds exhibit a unique yaw turn performance that sets them apart from other species such as fruit bats, hawkmoths, and fruit flies. While insects can also perform agile maneuvers, the more complex biomechanical characteristics of hummingbird wings provide significant advantages. These include greater control over wing kinematics, which leads to the generation of counter-torque, or torque in the opposite direction of the body's turn. As one of the few animals capable of executing agile yawing maneuvers, the hummingbird's flight aerodynamics serve as an excellent inspiration for drone design. The kinematics of the body and wings show that hummingbirds have pure yaw maneuvers with very minimal pitching, rolling, or vertical and horizontal movement of the body. The incidence, stroke, and twist angles of the inner wing (IW) are significantly greater than those of the outer wing (OW) in making it accomplish such fine-tuned control during maneuvers. Steady aerodynamic calculations indicate the drag-based asymmetric force created by both the downstroke and upstroke of a hummingbird's wings aids in controlling turn velocity, allowing fine steering maneuvers. The two loop vortices created during each half-stroke enhance the effect of precision maneuverability, as well. A wake analysis revealed that the asymmetric wing kinematics create a significant difference, around 59%, in leading-edge vortex (LEV) strength between the IW and the OW. Besides this, wing flexibility is crucial for generating the large torque required to complete turns and produce enough lift to sustain the bird's weight in these maneuvers [4]. Based on several features, modern UAV categories can be classified, such as aircraft type and the way lift is developed, like the traditional airplanes, where it produces through aerodynamic method, the helicopter type, with a method of lift from rotating blades of a propeller, the multi-rotor type utilizes more than one rotor to have stability in addition to providing lift [1, 5]. Biomimicry is now a major potential in the current market because most designs are inspired by nature. For instance, the MAV's design is primarily based on the way birds' flight mechanics operate. Engineers can then create these small and lightweight flying machines with high maneuverability and controllability. The engineering of the MAVs was inspired by the anatomy and behavior of birds in such a manner that flies more efficiently than the regular aircraft. These bio-inspired designs provide better performance in agility, stability, and energy efficiency, making them extremely beneficial for different applications [6].

9.3 Relevance of Hummingbird-Inspired Drones

In many manufacturing industries, robots are used to repeat similar tasks, and their construction usually relies on metal parts. However, nature provides a different way. Mechanisms in nature, which are bio-inspired, are usually made from soft materials. From this aspect, one can learn about designing and developing robotic structures by understanding how animals navigate and adapt to unpredictable environments. Such soft and flexible components inspire these natural designs to encourage more versatility and resilience in a robot. Such robots can, therefore, be considered as much better at handling more complex tasks and thriving conditions like those of living creatures in the wild.

The growing understanding of soft materials has pushed for developments in soft robotic structures inspired by animals such as worms, caterpillars, and even octopuses, used in flexible, adaptive robotic structures in complex, confined settings. However, actuation systems are the major concern in soft robotics. Those systems need to be as soft as possible, strong, and capable of transmitting considerable force. This challenge must be overcome to enable soft robots, such as these prototypes, to accomplish strength and dexterity tasks that real-world environments will require of them [7]. Soft robots use three key actuation methods. The three methods include DEA deformation through electric fields, SMA deformation based on temperature variation, and compressed air or pressure fluids, which enable bending motions. These methods offer a great possibility of performing highly complicated tasks with flexibility.

9.4 Evolution of Bio-Inspired Drones

Commercial aircraft are constructed with rigid wings on which engines provide power, and various sensor and control systems guide them in the right direction and orientation. Instead, birds are provided with flexible wings that facilitate dynamic movement; they can cover long distances without rest. The most of these natural capabilities get picked by bio-inspired wing design into three types: non-deformable wings, a shape similar to bird wings designed to ensure effective flight is (A); deformable wings, with the help of which bird wings give their flexibility and adaptability, ensuring more maneuverable performance, and (B), complex structures that merge biological elements with morphing motions for optimal performance and flight agility [8].

9.5 Hummingbird Flight Mechanics

Many researchers, through experiments, showed that hummingbirds can vary the range of bird-specific kinematic parameters both with respect to different flight velocities and within individual wing beats. Such an ability enables hummingbirds to keep precise control and adaptability in flight, thus contributing to their exceptional agility and maneuverability. The unique flying pattern of hummingbirds is very interesting, such that they can easily hover, maneuver, and even fly backwards. Another is that hummingbird aerodynamics allows for lift production in each half of the stroke and achieves the characterized flight pattern [9]. Hovering, most generally, the ability to generate the vertical forces, represents only one component of flight performance. Other axial forces underlying thrust generation are not studied in the hummingbirds [10]. Hummingbird-inspired drones can be smaller and more lightweight than conventional drones. They can take up less space and can be useful in situations where the drone needs to operate in small or crowded environments, like under trees,

Figure 9.1 The manageability and stability of a hummingbird.

inside buildings, or among dense obstacles. The manageability and stability of a bird are shown in Figure 9.1.

Adaptability to Small-scale Tasks: Their size and agility make them suitable for tasks like aerial inspection, surveillance, wildlife observation, or delivering small packages in highly confined spaces where larger drones might be impractical.

Hummingbirds produce a mass-specific power by their flight muscles at zero flight speed. Hummingbirds also produce power aerobically and can hover for an infinitely long time, which is another specialty. Due to the necessity of adaptation to different situations by fruit flies and hawkmoths, this leads to differences in flight mechanisms, such as clap and flying in wasps, and delays in stall and LEV.

9.5.1 Calibration of Sensors and Error Correction

Unmanned aerial vehicle platforms with lightweight sensors are often plagued by severe geometric and spectro/radiometric limitations. To remove these problems, self-calibration and pre-processing stages are necessary. Radiometric calibration, which includes color correction, noise removal, and deblurring, requires spectral targets with known reflectance properties. However, adverse weathering such as rain or wind can interfere with this by introducing spectral distortions in variable illumination, altered reflectivity, and partial absorption by materials.

Further distortion of the camera's basic parameters, like focal length and distortion coefficients, during a flight of UAVs due to their rapid maneuvers and frequent altitude and orientation changes can be removed using the geometric calibration process, commonly termed as orthorectification. Even though such a process is essential in ascertaining the accuracy of data collected, its procedure continues to be an area of continuous and challenging research.

9.5.2 Flight Time and Path Planning

Environment monitoring is limited by the flight time of drones, which can be overcome by optimized path planning and advanced battery technologies. Path-planning strategies take into account the area to be covered and UAV energy constraints, balancing coverage, accuracy, and duration. Research shows that simpler paths often require more energy due to sudden maneuvers. Advances in lightweight batteries, including solar panels, can also be used to increase flight durations from 1 to 5 hours. While UAVs are effective for small areas (up to 0.2 km^2), larger areas require complementary technologies for data collection.

9.5.3 Localization and Tracking

An effective monitoring of the environment requires accurate tracking of UAV position and orientation, from initial planning to post-flight analysis. Ground control stations rely on a precise prediction of trajectory to manage UAV swarms and avoid collisions. Accurate georeferencing linking images to physical locations using ground control points (GCPs) is the key to aerial photogrammetry. Unmanned aerial vehicles generally depend on onboard global navigation satellite systems and inertial measurement systems for positioning. However, in remote or obstructed locations where global navigation satellite systems are unavailable, these systems can be unreliable, causing trouble in accurate monitoring.

9.6 Characteristics of Hummingbirds

A detailed study conducted to analyze the wing deformation reveals some important results that describe twist angle increase and the hover span during flight, which offer greater input into the dynamics of hummingbird flight [3, 11]. The tests, like digital particle image velocimetry, were conducted to measure the wake characteristics of hovering hummingbirds [12]. These results showed clear evidence of LEVs over the wing surface, which is a well-documented mechanism for lift production in insect flight [2, 13]. The aerodynamic efficiency of hovering was further supported by the strong downwash patterns that occurred during the downstroke and the mean circulation generated during the downstroke, which was roughly 2.1 times greater than during the upstroke.

Many researchers used digital particle image velocimetry to capture the horizontal flow in hovering Anna's hummingbirds, revealing unique vortex loops underneath the wings that added greatly to the knowledge about flow in hovering flight [14]. Numerical simulations of ruby-throated hummingbirds in hovering flight at a Reynolds number of 3000 also showed that lift production during the downstroke was about 1.5 times greater than during the upstroke, attributed to the high wing angle of attack, drag-based force, and wing–wake interaction [2, 15]. Furthermore, it was found that the pitch reversal of the hummingbird wing is driven by wing inertia, without increasing power requirements, thanks to elastic energy storage at the shoulder joint [3, 15].

9.7 How Do Hovering Hummingbirds Work

Hummingbirds generate 75% of their weight support during the downstroke and only 25% during the upstroke. This asymmetry is likely due to the inversion of their cambered wings during the upstroke. The wake patterns of hummingbird wings reveal the presence of LEVs formed during the downstroke, suggesting that hummingbirds operate at Reynolds numbers low enough to exploit a lift mechanism typical of insect hovering [8, 10, 11, 16]. Although similar in some respects to insect hovering, hummingbird hovering is unique because of the peculiar effects imparted by their body structure as avian animals.

Flapping wings can potentially produce energy efficiency and excellent maneuverability, thus well-suited for small drones within a centimeter to decimeter size range. In such sizes, fixed wings do not provide enough lift at larger speeds, while rotor-based flyers are not very efficient in terms of energy production. Flapping wings, therefore, improve maneuverability within congested spaces and represent one of the biggest advantages of small-scale drones.

9.8 Materials and Fabrication Techniques

Artificial birds, in contrast, make use of DC motors for the movement of their wings. Three reasons for the choice of DC motors are as follows: they are widely available; they can be powered directly from the battery; and they supply enough power to lift the robots. The main challenge is selecting an efficient motor that offers high power with minimal weight.

Interestingly, successful hovering man-made flyers such as Colibri, Nano-hummingbird, Maryland-hummingbird, and KUBeetle have motor weight distributions that are even lower than those of average natural hummingbirds. This implies that these artificial flyers have highly efficient motors. In contrast, the Katholieke Universiteit Leuven (KUL)-hummingbird and Purdue-hummingbird have higher motor weight distributions, which limit the design of other components due to weight and lift constraints. Achieving a reasonable weight distribution is essential for designing effective flapping-wing micro air vehicles (FWMAVs). Notably, natural hummingbirds show impressive load-lifting capabilities, generating lift approximately three times their body weight on average. Hummingbirds fly due to the flapping action of their muscles and bones. In FWMAVs, this flapping is achieved through mechanical designs. Some notable designs are the dual serial four-bar linkage, as well as the stroke-cam flapping mechanism [17]. Here, it is shown that, regardless of the imperfections and uncertainties embedded within these mechanical designs, with proper active control systems, an FWMAV can remain stable and efficient in flight.

The Maryland-hummingbird uses an innovative flapping mechanism based on a five-bar linkage. In this design, the four-bar linkage powered by a DC motor initiates the flapping motion, and the addition of a fifth bar amplifies the motion to achieve the desired flapping amplitude. In the KUBeetle, two interconnected mechanisms mounted on a large pulley system are used to convert motion into wing flapping. A belt pulley is used to increase the amplitude of a flap. The wings can be brought to the required angle, and the overall flight performance is improved in this way. The weights of flapping mechanisms for successful hovering hummingbird-like robots are between 13%–14%. (At present, Colibri, Nano-hummingbird, Maryland-hummingbird, and KUBeetle are successful hovering flights.)

In Nan's study, a tail inspired by the hummingbird's tail was designed using 1 mm-thick balsa wood, with a maximum length of 70 mm. These handmade tails were found to help stabilize posture by generating a moment, a function similar to the findings in. This design principle can be applied to FWMADs to enhance stability by adjusting the position of the tail. While earlier studies involving twin-wing hovering FWMADs often utilized the application of wing twist modulators that introduce pitch and roll moments, newer studies have established that tails can passively stabilize FWMADs [18], alternatively to ensure balance during the flying process.

Size significantly impacts flight capabilities. Larger birds, unlike small insects such as hawkmoths and mosquitoes, struggle to hover. To hover, a flyer needs large enough wings to generate sufficient lift. Insects have light wings, making up only 0.35%–6% of their body mass (B.M.), such as a 19-mg crane fly with 0.6 mg wings [19]. Heavier flyers like hummingbirds and bats require larger wings, which account for 6%–12.9% of their body mass [20]. For example, a hummingbird, weighing 5.1 g, has wings weighing 1.16 g and a 117-mm wingspan. Flight muscles in insects comprise 17%–56% of their body mass but decrease in larger flyers, as structural mass increases to support their bodies. Nonetheless, larger flight muscles enable the generation of higher lift with bigger wings. [21] Wing morphology plays a vital role in understanding how natural flyers achieve efficient flight. Larger birds tend to have proportionally larger wing surfaces and longer wings to support their increased body mass. However, their wing beats slow down as body size increases, due to the energy and aerodynamic efficiency required for larger flyers. Typically, wing loading (weight divided by wing area) rises with size, but hummingbirds are a notable exception, maintaining low wing loading for hovering. Interestingly, the aspect ratio of wings remains similar across species, reflecting a universal aerodynamic balance. Because of the smaller size of the hummingbird, the effect due to mass considerations is negligible.

9.9 Application and Potential Uses

9.9.1 Surveillance and Reconnaissance

The merits of particle swarm optimization (PSO) and Penguin Search Optimization Algorithm (PeSOA) are combined to give good surveillance and reconnaissance. The search process is guided using PSO with particles representing UAVs. For improving the efficiency of the search, the attributes of PeSOA, including area-based group search and inter-group UAV movement, are added to improve the objectives of reconnaissance. This hybrid approach combines the flexibility of PSO and the cooperative strategies of PeSOA to achieve optimal UAV deployment. The area-based search follows the concept of holes and positions of PeSOA, while the inter-group movement of UAVs mimics the movements of penguins in response to the presence of fish. Each UAV group is led by a head drone, which communicates data—such as location and targets identified—to the central command, allowing information to be shared across all groups for coordinated reconnaissance efforts [22].

9.9.2 Medical and Resource Operations

Transportation of critical medical supplies by drones has revolutionized healthcare delivery. Drones can quickly deliver items, like vaccines, blood, and medications, to places like remote villages or areas affected by disasters, which are hard to reach. During emergencies, drones ensure life-saving supplies arrive exactly when they're needed by cutting through barriers like poor roads or traffic. By providing better healthcare facilities in underserved regions, this approach strengthens global health systems, ultimately reducing healthcare inequalities and saving lives [23].

9.9.3 Environmental and Ethical Considerations

Collisions, disturbance to wildlife, noise pollution, privacy violations, and data security are a few environmental and ethical considerations of drones [24].

9.9.4 Unmanned Aerial Vehicles for Air Monitoring

Unmanned aerial vehicles have opened new frontiers in air pollution monitoring, enhancing aerial inspection capabilities. Experiments in urban areas have shown that atmospheric aerosols and gases disperse differently at various elevations, highlighting the inadequacy of ground-level monitoring (1–5 meters) for accurate air quality assessments. Normally, this is overcome by UAVs that are equipped with gas sensors, which can monitor pollutants at different altitudes, providing richer, real-time data that improves long-term air quality analysis [25]. The high spatial and temporal sensing capabilities of UAVs also enable the creation of detailed air quality index maps, offering near real-time pollution profiling in urban and suburban areas. When integrated with plume models, like the Gaussian model, these maps help to better understand the dispersion of pollutants in the air.

Miniaturized infrared thermal cameras are also part of UAVs, which help in microthermal environmental monitoring by giving a high-resolution analysis of land surface temperatures and variations at the microscopic level [26]. This information, along with humidity, solar radiation, and wind speed data, supports critical services such as weather forecasting and assessing the chemical composition of clouds, especially in regions affected by volcanic activity or chemical disasters.

More recently, research has been conducted on the possibility of using UAVs for spectrum sensing, either spatially, temporally, or spatio-temporally, to detect and locate sources of electromagnetic pollution. Unmanned aerial vehicle systems with microphone arrays are also being researched for acoustic noise monitoring in urban environments. These are still in their infancy, though, as issues such as propeller-induced wideband noise and engine noise complicate accurate detection.

9.9.5 Unmanned Aerial Vehicles for Land Monitoring

Advances in miniaturized RGB cameras and multispectral/hyperspectral imaging have made UAVs an effective and cost-saving substitute to satellite and airborne remote sensing data, which can provide high-resolution data for soil monitoring and environmental applications. They are widely applied nowadays for illegal waste dumping, landfill management, soil contamination prevention, and unauthorized construction [27].

Unmanned aerial vehicles play a crucial role in natural disaster monitoring by estimating damage caused by wildfires, landslides, and tornadoes, thereby facilitating the quick production of high-resolution land-use maps for emergency response. They also support precision agriculture by providing field-level crop health and soil quality analysis, thus enhancing early diagnosis and lowering costs and environmental impact. Along with models such as the soil moisture analytical relationship, UAVs also enhance soil moisture monitoring, aiding more informed agricultural decisions.

9.9.6 Unmanned Aerial Vehicles for Marine and Water Monitoring

Hyperspectral cameras can capture the unique reflectance characteristics of water surfaces, which enables the detection of suspended solids, particulate matter, and toxic agents. Machine learning tools, including support vector machines, are very effective for detecting oil spills in optical images. Low-cost UAVs equipped with spectral cameras have been used to monitor sediment levels in natural reservoirs and assess submerged vegetation and algae species, which are indicators of water quality. Integrating UAV-collected water data with hydrological models enhances the accuracy of monitoring river, lake, and sea dynamics, improving the monitoring of critical phenomena like flooding and inundations [28].

9.10 Regulations and Policy for Operating Unmanned Aerial Vehicles

Drone operations are subject to the rules and regulations made by the government of a country, and these rules and regulations differ from country to country. The rules and regulations are made to protect other aerial animals that share the same flight environment and also to protect people and property, especially in case of damage to UAV platforms [29]. To date, the governments have not allowed operations to be carried out by an autonomous drone. A licensed drone operator is required even for the most basic operations. Because of the restriction to keep a drone operator, the minimum distance at which a UAV platform can sense environmental data is limited, and this is one of the greatest hurdles in deploying autonomous drones for environmental purposes.

9.11 Limitations and Challenges

Technical constraints, such as battery capacity, cargo capacity, and dependence on weather conditions, are the most frequently cited reasons that drones face while being used for the transportation of medical supplies. Due to adverse weather conditions like strong winds and rainfall, drones may not be available to deliver emergency supplies, making them unreliable to use during such situations. Battery capacity limits the range that these drones can travel [30]. They need to stop at charging stations or at a place where their batteries can be replaced if they must be used for an extended duration of time.

Inadequate policies regarding the operation of drones in some countries severely limit the potential impact of these drones. A lack of policies regarding drone operations while transporting hazardous biomedical goods is even more concerning. Many studies have consistently highlighted the need to develop regulations and frameworks focusing on noise and privacy. The use of UAVs for transporting goods has received very little attention from governments, industries, regulatory bodies, and academic institutions.

9.12 Current Challenges and Solutions

9.12.1 Manufacturing Flapping Wing Mechanism

Flapping wing mechanisms in bio-inspired drones use crank and rocker mechanisms. Though effective, this mechanism has a drawback as it does not allow adjustable strokes, limiting flexibility in flight control [31]. Other challenges, like manufacturing tolerances, motors with insufficient power density, and limited control systems, complicate the design of efficient bio-inspired drones.

A critical challenge is the miniaturization of components while maintaining sufficient control capabilities. For example, even a surveillance drone needs to carry at least a camera and a dozen other sensors to perform its intended use [32]. Therefore, a trade-off needs to be made between efficiency, miniaturization, and the complexity of control systems. Another challenge is the failure of these drones during operations. Failures not only cause financial losses but also compromise the success of the mission. Therefore, drones need to be reliable and robust in both design and operation to achieve the overall effectiveness of bio-inspired drones in these applications [33].

9.12.2 Grasping During Flight

Grasping and carrying heavy and irregularly shaped objects present significant challenges for drones, particularly small unmanned aerial vehicles (SUAVs). A topic that has garnered extensive research attention, in this context, is flight stability, which is of primary concern. Typically, SUAVs operate in lower airspace, where they are more prone to flight instability. Natural environmental conditions, such as wind, and urban structures, create unpredictable air currents that can give rise to air turbulence that causes this instability. When burdened with irregularly shaped heavy cargo, the drone's ability to maintain a steady flight path is greatly affected by these factors. Advanced control systems and robust design strategies are required to address these challenges [34]. These systems must compensate for the external influences of turbulence to ensure that SUAVs can carry out their tasks safely and efficiently. The utility of SUAVs in various applications, from logistics to search and rescue operations, can be significantly enhanced by the application of effective solutions in this area.

Inspired by the ability of birds and insects to adjust their posture to navigate turbulent conditions, researchers are exploring the possibility of bio-inspired flight systems to improve drone stability and agility. For better maneuverability in dense forests and faster cruising speeds, conventional drones, which have rigid structures, can benefit from adaptive shape variations [35]. At Delft University of Technology, researchers developed Delfly, a flying robot inspired by fruit flies. Delfly, weighing 28 grams, can accelerate up to 25 km/h by flapping its four wings and recover quickly from collisions. However, its flight time is limited to 6–9 minutes. Guido de Croon noted the challenge of creating autonomous flapping drones with minimal sensors and computing power. Researchers developed Bat Bot B2, inspired by bat wings, at the University of Illinois and the California Institute of Technology. Bat Bot B2 uses a thin silicon film that deforms during flight, enabling energy-efficient autonomous flight.

Researchers developed the X-wing ornithopter, inspired by swallow flight, at National Chiao Tung University. This ornithopter can fly at high speeds, perform aerobatic turns, hover, and dive by adjusting its tail for thrust [36]. These bio-inspired innovations aim to enhance drone performance by improving stability, energy efficiency, and agility. A comprehensive review by researchers at Chiba University on flapping-wing micro air vehicles in 2016 suggests that insects prioritize agility for survival, balancing between maneuverability and flight stability. Insects use multiple sensing systems to achieve quick and precise responses to constantly adjust their controls and stabilize flight. This mechanism can be used to make multiple-sensing and real-time control technologies for drones, to help them avoid collisions with birds and insects [37]. Current drones provide high agility with their four rotary wings using advanced control methods. Since the angle of attack of rotary wings is smaller than that of flapping wings, this makes the rotary wings less stable and maneuverable. Flapping wings can dynamically change their angle of attack in every stroke, thereby creating large wing motions that help avoid or delay stalling. Flapping wings also provide a large lift area, which augments torque and lift production during maneuvers, significantly improving the flight robustness.

Researchers at New Mexico State University, inspired by the high propulsive efficiency of collective bird flight, proposed a multi-flapping winged drone with pairs of V-shaped wings to utilize the aerodynamics of bird flight when they are flying in a V-shape formation. Researchers at Chiba University identified an optimal rotor configuration, inspired by insect flight, for a drone with four rotary wings that increased the lift force by 7%. Researchers at the Netherlands Organization for Applied Scientific Research proposed a nature-inspired algorithm to control SUAV swarms, reducing the need for extensive communication infrastructure and computational costs [38]. To replicate the extreme maneuvers in hummingbirds, a hybrid controller was created by combining a model-based non-linear controller with reinforcement learning.

These advances address key challenges such as avoiding collisions with other objects in the surroundings, grasping irregular and heavy-shaped objects, flight efficiency during long periods of operations, flight stability in complex environments and adverse conditions, and damage mitigation to aircraft and surroundings.

9.12.3 Damage Mitigation

The rotor blades of conventional SUAVs are designed for high aerodynamic performance, which makes them sharp and inflexible, posing a significant threat to people, animals, and property. Hence, it is essential to carefully select materials and structural designs for SUAVs. Researchers at EPFL, inspired by the flexibility and rigidity of insects, developed a dual rigid frame that softens and folds upon collisions to reduce damage. This innovation was successfully tested with a 51 g quadcopter, demonstrating a notable shock-absorbing effect [39]. This innovation was further improved by the same team, incorporating the flexibility of insect wings in an origami-like structure. The team mimicked the elastic protein found in insect wings by sandwiching an elastomeric membrane between rigid tiles. In this way, the wings not only softened upon collision to mitigate damage but also maintained their rigidity during flight.

Researchers at the Japan Advanced Institute of Science and Technology, inspired by the wings of dragonflies, developed a propeller with flexible nodules that can bend, twist, and absorb collision forces. It can return to its original shape within 0.4 seconds and continue its normal operations. The exceptional flight performance of dragonflies has inspired many studies on the aerodynamic design and control of dragonfly-inspired micro air vehicles [40]. Further research is required to make drones that can prevent damage to themselves and their surroundings. This ongoing research highlights the importance of bio-inspired innovations in enhancing the safety and efficiency of SUAVs in various environments.

9.12.4 Collision Avoidance

Small UAVs typically fly at an altitude of 1 km, where animals like birds and insects are also present. To safely navigate these environments, small UAVs need to prevent collisions with these animals. Indoor-flying UAVs also need to avoid collision with the walls, furniture, and other things kept in the surroundings. For this, a control system with advanced cognitive abilities is required. Flying robots can avoid collisions by using the mechanism of optical flow, by insects like honeybees. Optical flow is the velocity field of the image of the environment projected on the retina of the organism. Honeybees measure the distance between themselves and their surroundings using this mechanism. Researchers at Delft University of Technology utilized this mechanism by combining optical flow and other visual information, drones can avoid collisions by continuously adjusting their flight path using real-time data from visual inputs. With this information, drones can avoid collisions with objects from the front. The sensitive antennae of the insects can detect changes in airflow to avoid collisions with obstacles. Inspired by the aerodynamic imaging of mosquitoes, researchers at the Royal Veterinary College were successful in developing a drone on which differential pressure sensors were mounted. It could detect changes in airflow and avoid obstacles in the environment. Bionic Bird's MetaFly is a well-known biomimetic drone capable of flying indoors at low altitudes and avoiding collisions with obstacles. Its flight speed can be changed manually by changing its tail angle, and it can absorb damage upon collision. The company has also announced that its successor, the MetaBird, can be controlled by a smartphone. Other methods of avoiding collision are using event cameras, bioinspired sensors with microsecond reaction times, swarm intelligence for communication with other flying objects, and locust-inspired neurons for collision detection [41].

A limitation to avoiding collisions is that there are no standard obstacles. The shape, size, and velocity of obstacles keep changing, affecting the drone's required functionality. Therefore, it must make a list of minimum obstacles that a drone should avoid, ensuring compliance with the rules and regulations of various airspaces to safely navigate in diverse environments.

9.12.5 Flight Efficiency

Flight efficiency can be improved by using power-efficient components. It increases the duration of flight and is a straightforward way of increasing flight efficiency. Another way is to use biomimetic flight strategies to further enhance flight efficiency. Researchers at

New Mexico State University, inspired by the propulsive efficiency of the V-shaped formation of migratory birds, proposed a drone design with multiple flapping wings arranged in a V-shape. By testing various shapes and sizes of wings and airfoils, researchers could achieve a propulsive efficiency of 73.8% while flying forward with the help of this bio-inspired design.

Researchers at Chiba University, inspired by the fact that insects exhibit minimal correlation between their flapping wings and their bodies, identified an optimal rotor configuration for four rotor wings that increased the lift force by 7% compared to a basic rotor configuration. These biomimetic approaches, inspired by nature's efficient flight strategies, significantly enhance the performance and efficiency of UAVs, enabling longer and more energy-efficient flights. A German pneumatic manufacturing company called Festo has developed a biomimetic drone called Bionic Swift. This drone uses a system of plates to improve flight efficiency. When the wings go up, the plates open to reduce power consumption, and when the wings go down, the plates close to generate a large lift force. It also uses ultra-wideband communication technology to communicate with other aircraft to fly in a formation, mimicking the behavior of migratory birds, to further improve flight performance [42]. A straightforward way to improve flight efficiency is to reduce the aircraft's weight, but researchers at the Netherlands Organization for Applied Scientific Research proposed an algorithm, inspired by nature, to control SUAV swarms. In this way, SUAV swarms can collect real-time visual data for inspection efficiently. This decentralized approach is not only effective but also reduces the requirement of an extensive communication infrastructure significantly and lowers computational costs. By mimicking the collective behavior found in nature, SUAVs perform tasks more effectively while requiring fewer resources [43].

Another way to reduce aircraft weight is to externalize the power source and use technology like wireless power transmission. This technology can be used in insect-sized drones that are less than 10 cm in size and weigh less than 5 g. Researchers at the Toyota Central

R&D Labs developed an insect-sized flapping wing micro air vehicle that weighs just 1.8 g and achieves an untethered flight using electromagnetic wave transmission for power with the help of a high-frequency power receiver that boasts a power-to-weight density of 4900 W/kg, which is 5 times more than standard lithium polymer batteries.

This advancement allows for remarkable lift-to-power efficiency, measured at 3.6–3.7 gf/W during hovering flight, significantly outperforming previous insect-scale FWMAVs. For instance, Harvard University's RoboBee X-Wing, known for its stable hovering and maneuverability, has a lift-to-power efficiency of 2.4–3.0 gf/W. The Toyota Labs' FWMAV's enhanced efficiency and reduced weight represent a significant leap forward in the development of miniature, wireless-powered flying robots. Stop-of-flight or perching is another way of conserving energy. Researchers at the University of Pennsylvania [2] developed a drone with a vision-based system that enabled it to perch along randomly oriented lines in space, significantly extending its operational time.

At a small scale, magnetic forces dominate gravitational forces. Researchers at the City University of Hong Kong and Harvard University [2, 42] made use of this fact by attaching an insect-sized drone with small steel discs, which allows it to perch on vertical surfaces with the help of iterative learning control and an adaptive tracking controller, showing a practical use of magnetic forces on micro aerial vehicles. A novel approach to perching mechanisms involves biomimetic drones inspired by nature. Researchers at Beihang University [2, 43] developed a hitchhiking robot attached with adhesive plates that allow rapid attachment and detachment to and from various surfaces both in air and underwater. This innovative approach opens new possibilities for versatile energy-efficient perching mechanisms in micro aerial vehicles.

The small size of the drone limits the amount of weight it can carry. For example, insect-sized drones are too small to carry any object, whereas even for surveillance and reconnaissance purposes, a drone needs to carry at least a camera and other sensors. This necessitates the importance of achieving a balance between efficiency and the minimal weight that a drone must carry to fulfil its intended purpose.

Manufacturing flaws, motors with low power density, and limited control methods are some of the major challenges that prevent a bio-inspired drone from matching the capabilities of its natural counterpart. Increasing the degree of freedom of the flapping wings of a bioinspired drone necessitates the use of a sophisticated control system to sustain long-distance flights. These challenges can be overcome by using standardized designs and sophisticated control systems [44]. The smaller the drone, the more limited its functionality is. For example, even a drone required for surveillance missions needs to carry the weight of a camera and other sensors. Therefore, a trade-off needs to be made between the efficiency of the drone and the minimum amount of weight that the drone must carry to fulfil its intended purpose. Biomimetic drones need to have an advanced control technology to support their integrated design, but their level of control has not yet reached that of other robotic technologies like precise manipulators and agile mobile robots [22].

Most research in this field has focused on the development of structure and the dynamic analysis of flying animals. Researchers aim to understand the impact of body design on their flight performance when interacting dynamically with their environment. As researchers try to mimic their complex flying motion, the structure of the drone becomes intricate, requiring more advanced control methods [45]. Creating a drone as small as a hummingbird presents significant challenges due to the miniaturization of key components like actuators, motors with high power density, and a power source. Such components need to be incredibly lightweight and efficient at the same time to mimic the flight performance and capabilities of a hummingbird at such a small scale. Miniaturization of such components without sacrificing performance requires the use of advanced materials and innovative engineering solutions. Actuators and motors must provide enough power and precise control to enable the rapid flapping motion of the wings of the drone for flight. Power sources such as batteries must be sufficiently lightweight so as not to burden the drone with its own weight and still supply enough power for sustained flight. Integrating all these sensors and control systems into the tiny body of the drone is yet another challenge that needs to be overcome.

The trade-off between size, weight, and efficiency is critical. Every component must be optimized to ensure the drone can achieve stable and agile flight. Overcoming these challenges involves interdisciplinary efforts in materials science, electronics, and aerodynamics, and requires innovative approaches to design and manufacturing. Successfully developing a hummingbird-sized drone could revolutionize applications in areas such as surveillance, environmental monitoring, and medical delivery, showcasing the importance and potential impact of this engineering challenge.

9.13 Conclusions

While researchers have focused on addressing the yaw capabilities of hummingbirds through approaches like flapping platforms with passive mechanisms, the study on the intricate and flexible motions of hummingbird wings remains inadequate [3, 8]. With advancements in 3D printing parts, the invention of complex materials, and the availability of computerized modeling methods, the design and development of such devices has a lot of scope and opportunities. Most of the research focuses on the maneuvering abilities of insects and the forward and hovering flight mechanics of hummingbirds instead of the full range of flight mechanics of hummingbirds. This limitation prevents the development of a fully functional hummingbird-inspired drone with sophisticated flight capabilities. Biomimetic drones have been able to perform better than their conventional counterparts by mimicking the materials, mechanisms, and structures found in flying animals. These advances give great hope for the expansion of the areas of application of these drones. However, most research is focused on extracting specific features like morphology and flight mechanics of insects and birds, and applying them in drone design. Even after these advances, researchers have yet to replicate the highly integrated and embodied intelligence found in natural fliers.

Achieving such a system, which remains highly unknown, will require a novel approach to design. Research on insects suggests that they use a highly integrated feedback loop mechanism to govern their flight, where internal components, like muscles and the nervous system, interact with external structures like the exoskeleton and wings to create a dynamic feedback loop. Studies are also underway to develop an artificial intelligence system, inspired by the nervous system of insects, to design highly efficient control systems for autonomous drones. These capabilities can significantly enhance the performance of biomimetic drones and help them match the capabilities of their natural counterparts.

References

1. Ristroph, L. and Childress, S., Stable Hovering of a Jellyfish-like Flying Machine. *J. R. Soc. Interface*, 11, 92, 20130992, 2014.
2. Shyy, W., Berg, M., Ljungqvist, D., Flapping and Flexible Wings for Biological and Micro Air Vehicles. *Prog. Aerosp. Sci.*, 35, 5, 455–505, 1999.
3. Warrick, D.R., Tobalske, B.W., Powers, D.R., Aerodynamics of Hummingbird Flight. *Nature*, 435, 1094–1097, 2005.
4. Altshuler, D.L. and Dudley, R., The Ecological and Evolutionary Interface of Hummingbird Flight Physiology. *J. Exp. Biol.*, 205, 2325–2336, 2002.
5. Hedrick, T.L., Cheng, B., Deng, X., Wingbeat Time and the Scaling of Passive Rotational Damping in Flapping Flight. *Science*, 324, 5924, 252–255, 2009.
6. Henningsson, P. and Bomphrey, R.J., Span Efficiency in Gliding Flies. *J. R. Soc. Interface*, 10, 83, 20130099, 2013.
7. Ma, K.Y., Chirarattananon, P., Fuller, S.B., Wood, R.J., Controlled Flight of a Biologically Inspired, Insect-Scale Robot. *Science*, 340, 6132, 603–607, 2013.

8. Ravi, S., Crall, J.D., McNeilly, L., Gagliardi, S.F., Combes, S.A., Hummingbirds Control Hovering Position Using Both Visual and Mechanosensory Feedback. *J. R. Soc. Interface*, 13, 118, 20160602, 2016.
9. Lentink, D. and Dickinson, M.H., Rotational Lift: The Key to Hovering of Insect Flight. *Science*, 313, 5784, 1954–1957, 2009.
10. Nguyen, Q.V., Han, J., Bang, H., Design and Control of a Bio-Inspired Flapping Wing Micro Air Vehicle. *Int. J. Aeronaut. Space Sci.*, 17, 137–147, 2016.
11. Mohseni, K. and Mittal, R. (Eds.), *Bio-Inspired Fluid Mechanics*, Cambridge University Press, New York, 2014.
12. Orsag, M., Bogdan, S., Kovacic, Z.A., Survey of Unmanned Aerial Vehicles (UAV) for Environmental Monitoring. *Proc. 21st Mediterranean Conf. Control Autom*, pp. 1147–1153, 2013.
13. Tanaka, H. and Santhanakrishnan, A., The Mechanics of Hovering Flight in Nature and Its Application to Bio-Inspired MAVs: A Review. *Bioinspir. Biomim.*, 12, 5, 051001, 2017.
14. Ifju, P.G., Jenkins, D.A., Ettinger, S., Lian, Y., Shyy, W., Waszak, M., Flexible-Wing-Based Micro Air Vehicles. *AIAA J.*, 40, 11, 2125–2134, 2002.
15. Fuller, S.B. and Wood, R.J., Biologically Inspired Mechanisms for Flight Control in a Flapping-Wing Vehicle. *Bioinspir. Biomim.*, 5, 4, 045003, 2010.
16. Berman, S., Lindsey, Q., Sakar, M.S., Kumar, V., Pratt, S.C., Experimental Study and Modeling of Group Retrieval in Ants as an Approach to Collective Transport in Robotics. *Proc. IEEE*, 99, 9, 1470–1481, 2011.
17. Keennon, M., Klingebiel, K., Won, H., Andriukov, A., Development of the Nano Hummingbird: A Tailless Flapping Wing Micro Air Vehicle. *AIAA Aerospace Sciences Meeting, AIAA*, pp. 2012–0588, 2012.
18. Duan, H. and Li, P., *Bio-Inspired Computation in Unmanned Aerial Vehicles*. Studies in Computational Intelligence, vol. 505, Springer, Berlin, 2013.
19. Cheng, B., Deng, X., Hedrick, T.L., The Mechanics and Control of Pitching Maneuvers in a Hummingbird. *J. Exp. Biol.*, 214, 24, 4092–4102, 2011.

20. Sato, M. and Hagiwara, Y., Development of a Flapping-Wing Robot Inspired by Hummingbird. *Robotica*, 34, 12, 2777–2792, 2016.
21. Nakata, T., Liu, H., Tanaka, Y., Nishihashi, N., Wang, A., Sato, K., Aerodynamic Performance of a Bio-Inspired Flexible Wing with a Morphing Structure. *Bioinspiration Biomimetics*, 6, 4, 045002, 2011, https://doi.org/10.1088/1748-3182/6/4/045002.
22. Khalifa, H.K., Fuller, S.B., Wood, R.J., Insect-Inspired Vision-Based Obstacle Avoidance for Flapping-Wing Micro Air Vehicles. *IEEE Rob. Autom. Lett.*, 3, 3, 1832–1839, 2018.
23. Straw, A., Blanchard, M., Dickinson, M., Visual Control of Altitude in Flying Drosophila. *Curr. Biol.*, 20, 17, 1550–1556, 2010.
24. Zhao, L., Cheng, B., Deng, X., Aeromechanics of a Bio-Inspired Robotic Hummingbird. *Bioinspiration Biomimetics*, 15, 2, 025001, 2020.
25. Kato, S., Fujimoto, Y., Nakazawa, K., V-Formation Flying Using Multiple Flapping-Wing Drones Inspired by Migratory Birds. *Proc. 2020 IEEE/RSJ Int. Conf. Intell. Robots Syst*, pp. 3193–3198, 2020.
26. Festo AG & Co. KG, BionicSwift – Lightweight Flying Object With Collective Behaviour, 2021, [Online]. Available: https://www.festo.com/group/en/cms/13787.htm.
27. Zeng, Y., Song, X., Zhang, J., Liu, H., Bio-Inspired Rotor Configuration for Quadrotors With Enhanced Lift Force. *J. Bionic Eng.*, 18, 3, 600–611, 2021.
28. Sato, T., Kubo, T., Yamada, M., Wireless Power Transmission for Insect-Scale Aerial Robots. *IEEE Trans. Power Electron.*, 36, 6, 7073–7084, 2021.
29. Chirarattananon, P., Ma, K.Y., Wood, R.J., Cheng, R., Dynamics and flight control of a flapping-wing robotic insect in the presence of wind gusts. *Interface Focus*, 7, 20160080, 2017.
30. Chen, L.H.Y., Zhu, Y., Zhang, X., Xu, S., Bio-Inspired Adhesive Perching Mechanism for Micro Aerial Vehicles. *IEEE/ASME Trans. Mechatron.*, 26, 3, 1334–1343, 2021.
31. de Croon, G.C.H.E., De Wagter, C., Remes, B.D.W., Ruijsink, R., Mulder, M., Design, Aerodynamics, and Autonomy of the DelFly. *Bioinspiration Biomimetics*, 11, 4, 046003, 2016, https://doi.org/10.1088/1748-3190/11/4/046003.

32. Mintchev, S. and Floreano, D., Self-Folding Drone Wings for Crash Tolerance. *IEEE Rob. Autom. Lett.*, 1, 2, 776–783, 2016, https://doi.org/10.1109/LRA.2016.2529386.
33. Ramezani, A., Chung, S.-J., Hutchinson, S., Bat Bot (B2): A Biologically Inspired Flying Robot. *Sci. Robot.*, 2, 3, eaal2505, 2017, https://doi.org/10.1126/scirobotics.aal2505.
34. Nguyen, Q.V., Truong, Q.C., Chen, Y.-C., Design and Control of a Swallow-Inspired Ornithopter. *J. Bionic Eng.*, 17, 4, 699–710, 2020, https://doi.org/10.1007/s42235-020-0045-z.
35. Zhang, D. and Kovacs, J.M., The Application of Small Unmanned Aerial Systems for Precision Agriculture: A Review. *Precis. Agric.*, 13, 6, 693–712, 2012, https://doi.org/10.1007/s11119-012-9274-5.
36. Barry, A.J., Majumdar, A., Tedrake, R., Safe Navigation for UAVs in Dense Environments Using Monocular Vision. *IEEE Int. Conf. Rob. Autom. (ICRA)*, pp. 1478–1485, 2015.
37. Anderson, K. and Gaston, K.J., Lightweight Unmanned Aerial Vehicles Will Revolutionize Spatial Ecology. *Front. Ecol. Environ.*, 11, 3, 138–146, 2013, https://doi.org/10.1890/120150.
38. Watts, A.C., Ambrosia, V.G., Hinkley, E.A., Unmanned Aircraft Systems in Remote Sensing and Scientific Research: Classification and Considerations of Use. *Remote Sens.*, 4, 6, 1671–1692, 2012, https://doi.org/10.3390/rs4061671.
39. Linchant, J., Lisein, J., Semeki, J., Lejeune, P., Vermeulen, C., Are Unmanned Aircraft Systems (UASs) the Future of Wildlife Monitoring? A Review of Accomplishments and Challenges. *Mammal Rev.*, 45, 4, 239–252, 2015, https://doi.org/10.1111/mam.12046.
40. Mohammed, F., Idries, A., Mohamed, N., Al-Jaroodi, J., Jawhar, I., UAVs for Smart Cities: Opportunities and Challenges. *Proc. Int. Conf. Unmanned Aircraft Syst. (ICUAS)*, pp. 267–273, 2014, https://doi.org/10.1109/ICUAS.2014.6842278.
41. Sankar, S., Panwar, A., Iyer, N.R., Bio-Inspired Flight Dynamics for Micro Aerial Vehicles. *Curr. Sci.*, 113, 4, 599–608, 2017.
42. Floreano, D. and Wood, R.J., Science, Technology and the Future of Small Autonomous Drones. *Nature*, 521, 460–466, 2015, https://doi.org/10.1038/nature14542.
43. Ghadi, Y. and Tsourdos, A., Path Planning and Trajectory Optimization for UAVs: A Survey. *J. Intell. Rob. Syst.*, 84, 1, 1–21, 2016, https://doi.org/10.1007/s10846-016-0363-9.

44. Iosifidis, A. and Tefas, A., Sensor Calibration and Error Correction for UAV Vision Systems. *Sensors*, 16, 5, 705, 2016, https://doi.org/10.3390/s16050705.
45. Zufferey, J.C., Beyeler, A., Floreano, D., Autonomous Flight at Low Altitude Using Light Sensors and Little Computational Power. *Auton. Rob.*, 23, 3, 197–206, 2007, https://doi.org/10.1007/s10514-007-9052-3.

10

Biomimicry in Automotive Design: Nature-Inspired Innovations for Efficiency, Performance, and Sustainability

Abstract

Biomimicry is contributing to the revolution in vehicle design. In a quest for new automotive technologies, researchers are learning about everything from people to small animals. As a result, engineers are creating automobiles that not only incorporate revolutionary changes in the efficiency, performance, and sustainability of the automobile but also the beauty of nature. This ongoing search is expected to enhance the development of automobiles and foster the emergence of bio-inspired concepts and innovations. Enhancing the form, structure, and control of the car increases its efficiency and aerodynamics, which can assist in reducing emissions and fuel consumption. For instance, fish- and bird-inspired aerodynamic designs are used in lotus leaf green-inspired and self-cleaning efficient surfaces of automobiles. Besides, other advanced cooling improvement systems, such as those modeled similarly to termite mounds that enhance vehicle efficiency and effectiveness, are also being developed. There is a lot of room for bio-inspired developments in the dynamic field of biomimicry in the automotive sector. Due to the tremendous advances in research and technology, major automakers are progressively incorporating natural designs and mechanics into their automobiles. Universities and companies are also actively investigating and testing these ideas.

Keywords: Biomimetics, drag reduction, weight reduction

Jagadeesha T. and Sandip Kunar. Bio-Inspired Manufacturing Systems, (239–270) © 2026 Scrivener Publishing LLC

10.1 Introduction

Biomimicry is all about solving challenges in engineering by humans using inspirations drawn from the designs of nature, its systems, and processes. In the modern automobile industry, achieving efficiency and performance requires creative and innovative concepts. As evolution in strategies takes place, the focus on designs inspired by nature is growing rapidly. This chapter, on nature-inspired designs, divides the application of bio-inspired designs in automobiles into five categories: aerodynamic applications, structural applications, wheel assembly applications, powertrain applications, and digital manufacturing applications. Aerodynamic applications of bio-inspired designs improve the performance, fuel efficiency, and sustainability of vehicles. The approach of bio-inspired design is based on principles, strategies, and solutions from nature to solve complex design problems in the automotive industry. There are various designs inspired by nature that have been implemented for vehicle aerodynamics efficiency, such as streamlined body shapes and surface textures, which are discussed in this chapter. These findings provide useful information by reducing aerodynamic drag and improving stability conditions when driving heavy vehicles.

Looking into structural applications, for centuries, plants, animals, and even natural structures have come up with efficient and effective solutions to challenges such as strength, flexibility, and environmental stress. Biomimetics, the process of using natural systems as an idea source, has become a revolutionary concept in engineering and material science. The automotive sector has embraced the biomimetic strategy in developing materials that enhance performance and help minimize environmental impact. Another important application is in the wheel assembly components. Numerous studies have been conducted to enhance the tire profile using biomimetic features to reduce rolling resistance and boost tire grip. There are two tire manufacturing plants of this type. One of them copied the tree frog's toe pad to enhance performance on damp terrain, while the other one used the biology of the cheetah paw to increase tire friction.

The cheetah's paws are small and create little friction when running. Low energy consumption is therefore made possible.

Applications in powertrain components include the cooling system, which is crucial for sustainable power generation, and has increased focus on nature-inspired designs. For instance, a novel design for a cooling fan that effectively decreases noise levels. Additionally, a seal lip for a cooling pump in electric vehicles was developed using a biomimetic approach. This seal lip is essential for keeping the water-based coolant separated from the surrounding air. Also, as the changes to the automotive industry are shaken completely with digital manufacturing, technologies like additive manufacturing, generative design ensure increased production. One of the problems faced in digital manufacturing is that the patterns used do not provide both strength and resilience. To overcome this, one solution is to incorporate bio-inspired designs.

The future of nature-inspired automotive design looks promising. With continuous advancements in material science, renewable energy technologies, and artificial intelligence (AI), the potential for further integrating biomimicry into vehicle design is vast. As the global demand for sustainable, energy-efficient transportation grows, nature-inspired solutions to play an even more critical role in shaping the future of the automotive industry.

10.2 Classification of Bio-Inspired Designs in Automobiles

In this section, the first automobile designs and challenges are discussed. Afterwards, bio-design and manufacturing aspects are discussed.

10.2.1 Aerodynamic Drag Reduction Applications

For several decades, automotive engineers and designers have increasingly taken inspiration from nature to improve the performance, fuel efficiency, and sustainability of vehicles. The approach of bio-inspired design is based on principles, strategies, and solutions from nature to solve complex design problems in the automotive industry. There are various designs inspired by nature that have been implemented for vehicle aerodynamics efficiency, such as streamlined body shapes, surface textures, and active aerodynamic systems. Additionally, the environmental and economic benefits of these innovations and the potential for future development in the field are also being a study of interest.

Automobile aerodynamics plays a crucial role in determining a vehicle's efficiency, stability, and overall performance. To reduce fuel consumption, as well as emissions, and optimize the handling of the vehicle, engineers have, over long periods of time, sought to understand how an automobile can be designed more smoothly, considering aerodynamic aspects. One of the solutions evolved was that engineers began to borrow innovative principles from nature, designed specifically to solve fluid-dynamic problems. This form of design, known as biomimicry, comes from the observation and replication of biological systems and natural processes that have been developed over millions of years and are now being used with perfection. By nature, animals and plants have highly efficient ways of managing airflow, reducing drag, and enhancing energy efficiency. The automotive designer can thus develop solutions that are environmentally sustainable and technologically advanced by studying these animals and plants. Applications of nature-inspired designs into automotive aerodynamics are not only a testament to the innovation of biomimicry but also an acknowledgment of the need to strengthen technology with the natural world around us.

The various applications of bio-inspired design contribute to the development of automobile aerodynamics, including breakthroughs, advantages, problems, and prospects, including biomimicry and aerodynamics, streamlined Shapes, surface textures, and adaptive aerodynamics.

10.2.1.1 Biomimicry and Aerodynamics

Biomimicry is all about solving engineering problems by humans using inspirations drawn from the designs in nature. In the field of aerodynamics, several key principles from biology have been used in automobile design. These include the streamlined shapes of animals, surface textures that reduce drag, and mechanisms that adapt to changing conditions. A bio-mimetic flow control device known as an automatic moving deflector (AMD), inspired by the secondary feathers of a bird's wing, was used in an experiment to reduce drag in a three-dimensional vehicle. The aerodynamic performance of the AMD was assessed by measuring the drag force, surface pressure, and velocity field behind and close to the base surface of the designed body. Because the AMD was composed of lighter and stiffer materials, including hard straw board, it improved the performance of the intended body and considerably reduced drag. With this AMD, a maximum 19% drag reduction was attained by altering its longitudinal length.

The dynamics of AMD at the beginning of its operation demonstrated that it begins to manifest when the pressure in the narrow space between the slanted surface and AMD increases significantly more than that on the upper surface of AMD. Without AMD, a separation bubble formed above the slanted surface at a slant angle of 25°, and powerful longitudinal vortices that rotated counterclockwise formed along the lateral sides of the slanted surface. These flow structures resulted in high drag, creating low-pressure areas on the slanted surface, especially at the front edge and lateral edges. However, above the sloped surface, the AMD stopped the separating bubble from forming.

It is possible that the current device can be applied to automobiles. For instance, some passenger cars with fast backs, such as sedans with slant angles ranging from 20° to 27°, may be covered by the current AMD. The use of AMD may also provide some advantages for SUVs and hatchbacks with comparable slant angles. Its uses might be expanded to include some components of fluid machinery with significant flow separation, including the combustor and diffuser. To reduce drag in large trucks, a study was conducted on new types of bio-mimetic cab-roof fairings (CRFs), which are air deflectors positioned on the top of the cab. The 15-ton and 5-ton models' drag coefficients were significantly decreased by the sophisticated bio-inspired CRF by 20% and 22.4%, respectively. For the 15-ton and 5-ton versions, the side force also decreased by up to 8% and 9%, respectively, at a yaw angle of $\phi = 3°$. To clarify such a drag reduction process, flow properties (such as mean velocity field and turbulent kinetic energy) surrounding the forebody of vehicle models with and without CRFs were carefully examined using a PIV technique. These findings provided useful information for enhancing the design of new types of forebody designs by improving driving stability and reducing aerodynamic drag.

10.2.1.2 Streamlined Shapes

One of the most famous examples of nature-inspired aerodynamic design is streamlining. Designing the shape of an object allows it to move through air or water with little resistance. One of the most famous examples of streamlining is the comparison between modern car designs and bird and fish body shapes. For instance, the beak of a kingfisher bird is pointed and sharply tapered, reducing drag when diving into water, which inspires the design of the nose section in high-speed trains, such as the Shinkansen. A dolphin also displays a fusiform shape of body, which makes it swim in the water efficiently without much turbulence. As such, the shapes of dolphins have sparked research on cars that are made more tapered, rounded on the front, and smooth with no drag for better efficiency.

The biomimetic study on models of the peregrine falcon, belted kingfisher, sailfish, and dolphin optimized the high-speed train in three areas: the train nose shape, the cross-section shape, and the train nose type. The first task would be the creation of bio-models using computer-aided drawing software, followed by simulation for the airflow surrounding them to establish the drag coefficients under counterflow conditions. After that, a comparison and validation of the drag coefficients of each bio-model will be done. This means that bio-models with a projecting nose will have a greater drag coefficient since the stagnation point occurs more frequently in the case of a windscreen present.

These high-pressure regions around the windscreen are stagnation points in the head that increase the total pressure of the head, leaving the tail with a relatively lower pressure coefficient; therefore, increasing drag. It was found that with optimization criteria, a bio model can be designed in which the windscreen is removed and streamlined to reduce micro-pressure waves at the train head.

10.2.1.3 Surface Textures

Nature has also devised a solution to reduce drag through surface textures. There is an excellent example, which is shark skin that is composed of microscopic scales arranged to reduce friction and drag when the shark swims through water. Inspired by this, engineers have developed riblet films—microscopic ridges that copy the pattern found on shark skin. The riblets are applied to the surfaces of vehicles to reduce air drag and increase the efficiency of fuel usage. Similar textures have been tried upon by car manufacturers to reduce drag: dimples on golf balls are known to reduce drag by creating turbulent airflow over the surface. Applying such textures on vehicles boosts aerodynamics without compromising structural performance.

10.2.1.4 Adaptive Aerodynamics

In the natural world, creatures can often change the shape and movement of organisms to maximize the use of energy. For example, the gull changes its angle as it flies with its wings to maximize the lift effect while minimizing the drag force. In cars, such a principle has motivated the formulation of active aerodynamics. The aerodynamic systems, which include active adjustable spoilers and air intakes, can change their shape or position based on the speed or driving conditions to optimize airflow around the vehicle. Adaptive features enable manufacturers to enhance the aerodynamic performance of a vehicle without requiring constant adjustments by the driver.

One of the outstanding illustrations of nature-inspired aerodynamic design is the BMW hybrid sports car, the BMW i8. The aerodynamic system used by the i8 is innovative in design, having both passive and active attributes. The vehicle has been designed with smooth curved lines and surface textures to reduce drag; its inspiration was derived from streamlined marine animals. Additionally, the i8 incorporates active aerodynamic elements such as adjustable flaps and air vents that change based on speed, optimizing the car's performance in various driving conditions. Another example is Tesla's electric vehicles, such as the Model S, which showcase the benefits of aerodynamic efficiency in reducing energy consumption. Tesla has designed the Model S with a thin, low-drag shape inspired by nature's fluid movement, optimizing airflow to reduce resistance. The car has a streamlined underbody, and the back of the car is tapered like a fish. Tesla engineers have also made use of vortex generators

Table 10.1 Automotive application of bio-inspired design.

Bio inspiration	Function/ mechanism	Automotive application	Examples
Bird's wing	Flow separation and vortices are reduced	Adaptive morphing in design	Slanted surface of a vehicle
Bird's beak	Smooth flow with the least resistance	The nose design of the bullet train	Shinkansen
Dolphin body	Fusiform shape to reduce drag/ aero resistance	Tapered surface	Tesla vehicle
Shark Skin	Low-turbulent friction drags	Surface films	Coating for automobiles
Gull wing	Adaptive flaps	Aerodynamic components	Automobiles like BMW, Tesla, and Mercedes.

to control airflow around a vehicle, like some animals controlling air movement about their bodies. Table 10.1 shows the function and mechanism, along with the practical application of bio-inspired design.

The Mercedes-Benz Bionic Concept Car is also a nature-inspired vehicle. This car design is inspired by a marine creature, the boxfish. Its compact yet efficient shape inspires its usage, and from that very inspiration, Mercedes-Benz went ahead to produce a form so aerodynamic that it even lessened drag by 20% from that offered by conventional cars. The Bionic car's body structure is similar to the boxfish's angular and smooth shape, leading to a heavy reduction in air resistance and improved energy efficiency.

Nature-inspired automotive design offers several significant environmental and economic benefits.

Energy Efficiency: Saving in drag results in improved fuel performance, hence saving energy to allow such a vehicle to maintain desired speeds. This means, thereby, less fuel consumption and reduced carbon emissions. That is why it also remains very important in recent developments in the automotive sector in sustainability and reducing car emissions.

Cost Savings: Improved aerodynamic performance also reduces wear on engine components, which could lead to longer-lasting vehicles at very low maintenance costs. However, for electric vehicles, energy savings are much larger because drag reduction enhances a vehicle's range and total efficiency.

Sustainability: Automotive manufacturers can produce vehicles with a lesser environmental impact by using designs inspired by nature. Biomimetic solutions can reduce energy-intensive manufacturing processes, while natural forms tend to be more durable and resource-efficient.

Looking forward to the future, there are lots of potential innovations. Further innovations in materials science, 3D printing, and computational fluid dynamics will create even more efficient and effective nature-inspired aerodynamic features. Based on further animal and plant biomechanical research, new and creative design solutions may evolve. It can be concluded that nature-inspired design has proven to be a successful method for enhancing car aerodynamics. From streamlined body shapes to adaptive features and surface textures, biomimicry has provided solutions that have enhanced vehicle efficiency, reduced drag, and improved sustainability. The cases of BMW i8, Tesla Model S, and Mercedes-Benz Bionic concept car outweigh the promising applications of nature-inspired innovations. With time, the role of nature-inspired design will only continue to grow within the ever-evolving automotive industry. It presents new opportunities for creating technically advanced vehicles that are environmentally sustainable with high aerodynamic efficiency. By continuing to learn from nature's time-tested strategies, exciting potential can be realized for both performance and environmental impact in the future of automotive aerodynamics.

10.2.2 Structural Applications

Nature has always been a great source of ideas for technological advancement and sustainable solutions. For centuries, plants, animals, and even natural structures have come up with efficient and effective solutions to challenges such as strength, flexibility, and environmental stress. Biomimetics, the process of using these natural systems as a source of ideas, has become a revolutionary concept in engineering and material science. The automotive sector has embraced the biomimetic strategy in developing materials that enhance performance and, at the same time, help to minimize weight and environmental impact.

For instance, natural composites based on jute, flax, and pineapple leaf fibers are being used in the production of automobile parts, including dashboards, door panels, and mudguards, whereby lighter, stronger, and more environmentally friendly designs are achieved. Also, honeycomb structures found in beehives have been used to come up with lightweight and strong structures of automobile frames and crashworthy parts. The other one is the application of layered composites, which are similar to seashells and which incorporate nacre's toughness to create impact-resistant materials. Thus, by applying these bio-inspired designs, manufacturers can meet the objectives of the industry, including the reduction of vehicle weight to increase fuel efficiency and the achievement of the sustainability goals with the application of biodegradable materials. This paper explains how biomimetic principles are changing the face of automotive material science by presenting potential solutions to strength, durability, and environmental concerns.

The study of nature serves as a fountain of ideas from which new and advanced materials and structures can be developed. This part provides a review of some biological specimens that helped scientists in their material science endeavors and explains how they can be used in automobile designs. It is most likely the aesthetic appeal of the comb-like structure of bees' hives that revolutionized the design of materials in automobiles by creating a widely accepted pattern that is both functional and attractive. This structure consists of several hexagonal cells that offer tremendous strength while using little material and keeping weight to a bare minimum. It is highly commendable that weight, safety, fuel efficiency, and overall performance of automotive materials are of concern.

Besides the exceptional weight-to-strength ratio, the hexagonal design also allows for weight to be distributed uniformly across honeycomb structures, making them ideal in areas with stringent weight constraints. An important characteristic of honeycomb structures is energy absorption during impacts, which consequently enhances the impact resistance of safety-relevant components. Such structures can be made of metals, polymers, or composite materials, depending on the purpose. Carbon fiber honeycomb materials can be used effectively in the structural design of aircraft fuselages, as well as in residential buildings. In ordinary compound applications, honeycomb structures can also be integrated into crash zones to help the car absorb and dissipate energy in the event of a collision, thus providing extra protection to passengers and minimizing injuries.

For example, aluminum honeycomb cores help absorb and compress shock forces in bumpers and side collision panels.

The use of honeycomb structures has also increased in automotive heat shields and insulation materials. The geometric design is effective in providing thermal resistance and is thus used in managing heat from engines or batteries in electric vehicles. Current methods, such as additive manufacturing or 3D printing, enable the creation of honeycomb structures with lightweight alloys or composite materials. Other methods, such as adhesive bonding and sandwich construction, are also used to integrate honeycomb cores with outer layers of aluminum or carbon fiber for increased strength. Honeycomb-inspired designs are still evolving in automotive material engineering to meet diverse needs and challenges, including weight reduction, fuel efficiency, and safety, with design ideas learned from nature.

Nacre, otherwise known as mother of pearl, is a natural composite material that is obtained from the inner shell of mollusks and the outer coating of pearls. It has been one of the most frequently used examples in biomimetics due to its high strength, low weight, and high resistance to fracture. These properties are the basis for applying nacre in the automotive industry since it offers solutions to problems such as impact resistance, durability, and weight. Many car manufacturers nowadays incorporate lotus-effect coatings with their exterior paints to achieve a self-cleaning surface. Such coatings repel rainwater, mud, and dust and, therefore, ensure longer durations of cleanliness with a minimum amount of vehicle maintenance. Hydrophobic lotus-inspired coatings in higher-end vehicles offer protection for an extended duration against environmental pollutants, such as bird droppings, acid rain, and tree sap. Applications to car windshields and side mirrors by repelling rainwater during rain improve visibility in such inclement weather. This, of course, eliminates the use of windshield wipers and decreases dirt accumulation. Superhydrophobic glass coatings are now commercially available. It is a durable, water-repellent surface for safer driving, inspired by the lotus leaf. The lotus effect is employed in car interiors to provide stain-resistant and easy-to-clean seat surfaces, dashboards, and carpets. Such a surface would repel liquids such as coffee or soda, for instance, not leaving any blemishes, and would facilitate cleaning. Most luxury cars also employ lotus-mimicking materials in leather upholstery and floor mats for added durability and ease of maintenance. Anti-water and mud adhesion with lotus-effect coatings is done to the underbody components and metal surfaces to avoid rust and corrosion. Protective wheel wells and undercarriage coatings enhance vehicle durability in wet or snowy weather conditions.

The pattern on the surface of a lotus leaf is replicated by engineers using nanotechnology at the micro- and nanoscale. An example is applying nanostructured polymers or silica-based coatings to superhydrophobic properties of car surfaces. One of the technologies used in depositing thin hydrophobic material films on surfaces is chemical vapour deposition (CVD). This process facilitates uniform coatings that could closely mimic the waxy texture of the lotus leaves to ensure strong durability and water repellence. Superhydrophobic spray coatings and films can be directly applied to automobile surfaces. The products in this category are cost-effective with multi-functionality, thus finding applications in various components, including glass, paint, and fabrics. The fact that the vehicle does not need to be washed as frequently is a significant benefit of vehicles with self-cleaning coatings, as they remain polished longer. The frequency of cleaning and maintenance is reduced by the lotus-effect surfaces since water, dirt, and contaminants are repelled from such a surface. Hydrophobic coatings applied to windshields and mirrors improve visibility when driving in rain. Self-cleaning surfaces reduce water, detergents, and consumption of cleaning agents, thus fitting into the sustainability trend in the automotive industry.

Superhydrophobic coating can degrade through wear and environmental exposure with time; further investigation is being done to enhance durability. One of the limiting factors of lotus-inspired coatings applied to mass-market vehicles is that production and application are very expensive. Up until now, the industrial manufacturing of large homogeneous lotus-effect surfaces in vehicles has remained a challenge. Future developments in material science and nanotechnology will most likely overcome these challenges, enabling the broader application of lotus-effect materials. In doing so, added value will be given to vehicles through their performance and reduced environmental impact, while convenience and safety are brought to customers. Shark skin has been one of the great inspirations in developing advanced materials in a variety of industries, including automotive engineering. The surface of shark skin is covered with microscopic ridges called dermal denticles. These denticles provide drag resistance in water, protection against biofouling, and increased efficiency and agility for the shark. Translating these properties into automotive materials has driven innovation in aerodynamics, fuel efficiency, and surface durability.

Contrary to what one might think, the skin of a shark is not at all smooth, but instead covered with a layer of overlapping structures similar to teeth called dermal denticles. These are ridged and aligned in such a way that they channel water effectively along the body of the shark. The microstructure of the skin reduces turbulence and drag, thus allowing the shark to move fast with less energy expenditure. These ridges on shark skin prevent microorganisms from attaching themselves to the surface. This keeps sharks clean and low-maintenance. The tough and resilient structure of dermal denticles makes shark skin resistant to wear and environmental stress.

Shark skin-inspired designs by engineers are being used to enhance the aerodynamic properties of vehicles. The surface materials mimic the ribbed texture of the dermal denticles, reducing air resistance and improving fuel efficiency and vehicle speed. Micro-patterned coatings or films that reduce drag in car exteriors are designed, especially for high-performance sports cars, inspired by the efficiency of shark skin. Reduced drag would mean that vehicles would need less energy to overcome air resistance. All this becomes crucial with the newer electric and hybrid vehicles, where every bit saved on range becomes important. Certain electric cars use shark skin-effect films inside their bodies for range efficiency, mainly at higher speeds. Anti-fouling coatings, translating the property of shark skin to prevent the buildup of microorganisms, are replicated onto the vehicle exteriors. Such coatings repel dirt, water, and other contaminants, thus keeping the vehicle clean with minimal maintenance. Sharklet Technologies is a biomimetic surface company that has developed anti-fouling surface treatments, inspired by shark skin, under test for use on automotive paint to reduce grime accumulation.

Spider silk is a material that is both lightweight and ultra-strong, with a tensile strength comparable to steel. Its structural arrangement allows it to have energy dissipation under stress; hence, the perfect model for impact-resistant materials. Advanced polymers and composites have been able to produce suspension springs and dampers using spider silk's molecular inspiration. These newer materials reduce weight while maintaining superior durability and flexibility. The lightweight suspension springs used in high-performance vehicles are made from carbon fiber-reinforced polymers inspired by the silk structures. Leaves combine strength and flexibility with the most effective distribution of their weight. The pattern of veins gives structural integrity to a minimum amount of material. Composite leaf springs in vehicle suspension systems have been designed using the same structural efficiency as nature. These are glass-fiber-reinforced plastics that achieve substantial weight reductions from conventional steel springs. One such composite leaf spring is installed in the rear suspension of the Chevrolet Corvette. While biomimetic innovations hold huge promise, some challenges that must be surmounted include the scalability of nature-inspired complex structures, high costs related to advanced materials, and durability under real conditions. However, nanotechnology, additive manufacturing, and material science continue to bridge such gaps. For example, 3D printing allows one to fabricate geometries inspired by nature, as intricate as honeycomb structures, or patterned like shark skin, with unprecedented degrees of precision and efficiency. On the other hand, studies on continuous bio-inspired nanocomposites and coatings make those materials increasingly economically viable while their performance is improved.

Table 10.2 A summary of various bio-inspired applications in automobiles.

Type	Bio inspiration	Application	Advantages	Examples
Biomimicry	Bird feathers	Active flow control	Drag reduction up to 25%	Slant surface vehicles, such as the Tesla Model X
Body shapes	Dolphin and falcon	Nose design	Flow continuity	Shinkansen Train
Surface texture	Shark skin and the wings of a butterfly	Micro grooves	Fuel saving	Racing cars
Adaptive surfaces	Fish fins and owl neck motion	Cab roof	Side force reduction	Trucks used in the military
Morphing structures	Bird wings and beetles	Change of shape during flight	Real-time optimization	Tesla Active Spoiler
Vortex control	Boxfish	Underbody design and diffuser design	Stabilized flow	Mercedes-Benz Bionic car
Optimization cross-section	Sailfish	Carriages for laminar flow	Low-pressure drag	Sail vehicles
Thermo Aero combo	Termite mounds	Ventilation and temperature control	Thermal comfort	Electric vehicles with passive cooling
Texture-based design	Cactus, leaf	Nano coating	Self-cleaning	Paint for automobiles
Bioactive materials	Octopus	Color and texture-changing materials	Stimuli-responsive systems	BMW Gina Light Visionary Model.

Nature-inspired materials and design in automobiles are at their very inception and hold immense promise. Biomimetic solutions for lightweight composites and aerodynamic coatings will feature large in extending range and efficiency in future electric vehicles, with research already in the pipeline for nacre-inspired thermal barriers to lotus-leaf coatings for battery casings. Innovations in AI-driven design optimization and 3D printing will make nature-inspired designs more accessible and scalable for mass-market vehicles, rather than just luxury or high-performance models. Bio-mimicry will be right at the heart of cars going greener as carmakers increasingly focus on sustainability. Biomimetic materials, functionally emulating natural structures, offer not only performance benefits but also environmentally friendly production and recycling. Future development in biology, engineering, and material sciences will also be very innovative since their clue have often come from nature. A summary of applications, advantages, and examples of bio-inspired automobiles is shown in Table 10.2.

10.2.3 Wheel Assembly Applications

In a car, the components of the wheel assembly are responsible for managing the orientation and load of the tire. These parts must be strengthened to support heavier weights. As a result, the vehicle's net weight is significantly increased using these sturdy materials. The wheel assembly considers the wheels, hubs, uprights, suspension, and braking parts. An incredibly popular futuristic idea that has increased vehicle performance while lowering weight is non-pneumatic tires. Because it can withstand high stresses, the honeycomb structure has been utilized for these tires the most.

Improving ride comfort in the vehicle can be significantly achieved through the high deformation potential of tires. Even today, off-road vehicles use such non-pneumatic tires. Scores of biomimetic studies have been carried out to boost rolling resistance at the same time as enhancing the grip of tires with specific shapes. There are two types of such tires: one emulating the adhesive pad of tree frogs for better performance in wet areas, and the other with a pattern inspired by the paw of a cheetah to increase tire friction. Whereas the feet of the cheetah are relatively small, they create minimal friction when running. This enables low energy consumption. Upon reducing speed and changing direction, their feet enlarge, thereby making it more stable and frictional. Consequently, it makes reduced energy usage possible. Their paws enlarge as they slow down and change their racing orientation. Its stability and friction are increased by this method. The cheetah's paws are therefore designed for maximum stability in bends, effective direction changes, and acceleration. The dry tire uses the cheetah's strategy. It is as broad as a regular tire, but as you brake, it gets wider, cutting the stopping distance by as much as 10%. Another kind of tire can roll without slipping in damp conditions.

The model for this wet tire was a tree frog's toe pad microstructure. Tree frogs can remain attached to damp surfaces, and the microstructure of their toes is made up of hexagonal cells with tiny grooves separating them. When the frog is on a damp surface, the fluid flows through these grooves, creating a thin layer of fluid between the pad and the surface that causes the pad to adhere to it by wet adhesion. Nevertheless, some biomimicry-based improvements in the suspension components were discovered, such as a spring-damper system with a tendril structure modelled after the bio-climber plant.

The design of a smart spring damper with variable damping using magneto-rheological (MR) fluid is the principal enabling concept for the present endeavor. The spring consists of a hollow tube filled with the MR fluid. While this traverse coil winds the tube, the MR fluid's stiffness and damping constant can both be adjusted by changing the conditions of the electrical fluxes through the winding. Another biomimetic damper is multifunctional and adaptable to futuristic designs. The biological links of the abalone shell inspired this kind of damper. This is because a possible new method of energy release is through modifying the biomechanism of sacrificial connections and hidden length.

Moreover, reactions to bonding can be expected in this way: sacrificial bonds, which are weak, and reformable ones that break before seriously damaging the strong bonds, keep the complete structure intact. When every sacrificial bond breaks, the primary strong bond will have the flexibility to extend a longer length; this is the hidden length. Force-displacement curves resulting from the cooperation of several molecules in a distributed passive damping system seem to have a greater consistency. This produces a forcing behavior that can easily be incorporated into dampers for applications in structural control. Thus, this type is considered more efficient than the conventional method of using dashpots and springs in a conventional damper. The hydraulic components in the biomimetic damper produce quite considerable dampening forces. A small air quantity is allowed to enter the system to mimic the original damping properties obtained from sacrificial bonding.

The air bubble in the hydraulic fluid aids in reducing hydrostatic shock when the displacement is small and as near the equilibrium location as possible. Damping is due to the air compliance produced by a slow raising of the load. Negative displacements are hardly present in animal bones. This damper, however, was designed for both positive and negative displacement values. A brake disc was created by taking inspiration from the locust's surface and the owl's wings, which decreased noise and increased the disc's resistance to scratches. The study suggests a modeling platform that can map many biological functions and generate a conceptual design that adheres to both references to accomplish this optimal design.

Weight reduction is a key consideration when building a vehicle with good performance and fuel economy. Since the wheel assembly parts are constructed of heavy metal to support heavier loads, their weights are significantly higher in this case. Therefore, one of the most important requirements for an automobile is to have sturdy and lightweight structures. Using biomimetic reference models, topology optimization is frequently employed in automotive design to accomplish this. Additionally, biomimetic topology optimizations were included for the uprights, brake calipers, suspension arms, and brake paddles.

10.2.4 Powertrain Applications

One could consider a vehicle's powertrain to be its central component. The powertrain contains numerous elements that obstruct increased performance and fuel economy. This system is the most difficult to recreate from living things when it comes to ideas influenced by nature. Nevertheless, hydrogen-powered solar vehicles that use sunlight and water as fuel sources have become a major topic of research, based on the biological energy conversion principle. There are examples in the literature where biomimetic design techniques are used for important automotive components and subsystems, despite the rare use of biomimetic principles in power generation.

These include components like hulls, pumps, nozzles, heat exchangers, cooling fans, and rotary impellers. The cooling system, which is essential to the production of sustainable power, has drawn more attention to designs inspired by nature. Additionally, a biomimetic technique was used to design a seal lip for an electric vehicle's cooling pump. Maintaining the water-based coolant's separation from the ambient air requires this seal lip. The hydrated lubrication mechanism seen in natural articular cartilage has been applied to the seal lip in this design.

Some papers indicate that the bio-inspired seal lip has effectively separated the coolant from the atmospheric air, thus reducing shaft friction when compared to the traditional oil seals. Another study proposed a liquid cooling module based on the human respiratory system. The purpose behind this invention is to enhance thermos-hydraulic efficiency. The human respiratory system introduces oxygen and removes carbon dioxide, inducing blood circulation. The coolant absorbs heat through a different mechanism, which is also proposed here to circulate and dissipate heat. For this reason, the coolant passes through a copper foam plate similar in configuration to the alveoli of a human respiratory system for heat absorption.

As with many other biomimetic designs of heat-powered adsorption cooling systems used in car air conditioning, it has been recorded in the literature. This cooling mechanism generally works remarkably well in car air conditioning because it is usually thriving while waste heat is available. The use of honeycomb structures in adsorption bed designs has, by now, proven the remarkable benefits of low costs, low weights, and good heat transfer qualities. Additionally, the literature about diesel particle filters identifies the honeycomb type. These systems have been used to comply with soot emission standards.

10.2.5 Applications in Digital Manufacturing

The changing face of automotive is completely shaken because of digital manufacturing: applications of various advanced technologies, such as additive manufacturing, digital twins, generative design, and smart factory systems, always ensure increased production efficiency, sustainability, and design-performance innovations. Digital manufacturing integrates technologies like 3D printing, digital twins, and smart factories to redefine the way cars are designed, produced, and maintained.

Traditionally, infill patterns used in additive manufacturing are characterized by geometries like honeycomb or gyroid, among others. Such configurations, while efficiently reducing materials used, are still ineffectual in scaling the property balance sought in most advanced applications. For example, overuse of material can take away from weight savings that additive manufacturing provides; conversely, using too little material can make the component mechanically fragile and, hence, render it useless. Overcoming this temporal barrier, current bio-inspired designs may offer unique solutions due to nature's expertise in saving materials by optimizing structures. Bio-inspired design, the concept of mimicking complex problems of nature with engineering, has garnered popularity in several science and engineering fields. Strength-to-weight ratio, responsiveness to the environment, and adaptability are strong features of bird nests, cocoons, or micro-architectures found in nature. Because of these traits, they may act as a template to inform the development of advanced infill patterns in additive manufacturing. Incorporating biological principles can help realize two objectives: one, the use of less material utilized, and two, preservation or initiation of augmented mechanical properties for the manufactured components.

The present investigation entails further exploration of the development and testing of bio-inspired material architectures for additive manufacturing. A study on three bio-inspired templates of infill patterns: bird nest, cocoon, and sandwich structures, is conducted. This study systematically evaluates the performance of these patterns when subjected to three conditions: applied load, strength, and impact investigations, where experimental methods, such as computer-aided design, 3D printing, and experimental tests, were conducted.

To develop the mechanical advantages of bio-inspired infill patterns in additive manufacturing, researchers employed a systematic approach. Following four phases: bio-inspired design development, additive manufacturing, testing materials, and a comparison with classical infill patterns. The bio-inspired design process began with determining the natural systems demonstrating a range of attractive mechanical properties (strength, resilience, and lightweight). Through literature reviews and database inquiries, the researchers chose three: bird nest-inspired connections to branch-like arrangements. This design aims to spread applied loads efficiently, minimizing weight. Modeled after layers and the dual-structured architecture of cocoons, which merges a stiff internal layer with a very compliant outer layer. Mimicking biological materials, such as bone and tree trunks, in alternating layers of hard and soft materials to achieve strength along with flexibility.

In this work, experimental results from bio-inspired designs were compared with those from traditional infill patterns to ascertain their relative performance. Points of analysis included any trend in mechanical behavior, such as weight versus load-bearing capacity, and the nature of trade-offs between resilience and material usage. The critical observation indicates that bio-inspired designs offer superior weight-to-strength ratios and resilience compared to traditional patterns, as provided by optimal material distribution and structural efficiency of bio-inspired patterns. Adopting bio-inspired principles, progressive computer-aided design packages, and systemic experimental methods allowed researchers to produce pioneering infill patterns to tackle some of the biggest issues in additive manufacturing. This exhaustive method supplies a pathway into further investigation of bio-inspired material architectures in various engineering sectors.

The study underscores the key transformational role bio-inspired design can play in elevating additive manufacturing strategies. Careful execution of tests and analysis led these engineers to conclude that natural systems may provide superior functional benefits for infill patterns over traditional designs in key metrics such as strength-to-weight ratio and resilience. Bird-nest and cocoon patterns showed high performance in terms of mechanical functionalities, coupled with low material requirements, meeting a unique challenge in additive manufacturing. Sandwich-inspired structures would serve in fields requiring high resilience and load capacity, such as aerospace and structural engineering, while bird nest and cocoon designs are more appropriate in lightweight applications. Bio-inspired principles provide engineers with insight, enabling them to create highly optimized material arrangements to balance functionality with sustainability.

10.3 Challenges in Adopting Bio-Inspired Designs for Automobiles

Adopting bio-inspired designs in automobiles presents several challenges, despite the potential for innovation in sustainability, efficiency, and functionality. Some of the key challenges include:

10.3.1 Complexity of Nature's Designs

Understanding Biological Systems: Nature has evolved highly specialized and complex systems over millions of years. Translating these systems into engineering solutions can be extremely difficult. Bio-inspired designs often require a deep understanding of biological processes, which may not directly translate into simple, scalable engineering solutions. Biomimicry Limits: While nature offers incredible efficiency, not all biological concepts apply to mechanical and electronic systems in automobiles. For instance, an efficient energy mechanism in an animal may not have a direct counterpart in automotive power systems.

10.3.2 Material Limitations

Availability and Cost: Many bio-inspired materials are not yet readily available or cost-effective. While natural materials, like plant fibers or certain minerals, might offer advantages in weight reduction or energy efficiency, they may be difficult or expensive to manufacture in bulk or integrate into vehicle design. Durability and Scalability: Biological materials often degrade under conditions such as UV light or temperature extremes that automobiles face. Reproducing the durability of natural materials without compromising environmental sustainability is a key hurdle.

10.3.3 Technological and Engineering Constraints

Integration with Existing Technologies: Bio-inspired designs might not easily integrate with current automotive technologies and manufacturing processes. Retrofitting current infrastructure to accommodate novel bio-inspired designs may be costly and complex. Performance Expectations: Bio-inspired solutions are often optimized for their specific biological context, not necessarily for high-speed, high-stress environments like automobiles. For instance, while aerodynamics in nature is designed for stability and efficiency in specific contexts, it may not always be suitable for automotive speeds or crash protection.

10.3.4 Manufacturing and Production

Unlike traditional designs, bio-inspired structures may require highly specialized, bespoke manufacturing techniques. This can increase production costs and complexity, especially when dealing with biomimetic components that have unique geometries or material properties. Lack of Standardization: Many bio-inspired designs involve innovative materials or structures that don't fit neatly into the existing regulatory and manufacturing standards for vehicles. Developing new standards and regulations to accommodate these designs can be time-consuming and costly.

10.3.5 Cost of Research and Development

Developing bio-inspired designs often involves significant research and testing to validate their performance, safety, and reliability. For example, mimicking the skin of sharks for reducing drag in vehicles would require detailed studies, prototypes, and simulations before it could be applied in production. Uncertainty in Return on Investment: Investing in bio-inspired technology can be risky for automakers, as the long-term benefits might not always be immediately apparent, and the cost of research, development, and integration could outweigh the potential gains, especially if the designs do not perform as expected.

10.3.6 Sustainability Concerns

While bio-inspired designs aim to increase sustainability, the process of scaling these designs for mass production may introduce its own environmental costs. For example, some natural materials that seem sustainable in their original form may require energy-intensive processes for manufacturing, transportation, or disposal. Energy and Resource Efficiency: While bio-inspired designs often aim to be more energy-efficient or resource-conserving, they may face challenges when scaled to meet the performance demands of modern automobiles, especially in areas such as engine design, safety features, and longevity.

10.3.7 Regulatory and Safety Compliance

Bio-inspired designs may not yet be fully understood by regulators, leading to challenges in getting new materials or systems approved for use in vehicles. Safety regulations, crash test standards, and environmental impact assessments would all need to be revisited to accommodate bio-inspired technologies. Uncertainty in Testing Protocols: Standardized testing procedures may not account for the unique behavior of bio-inspired materials or systems, leading to ambiguity in determining whether these technologies are safe or effective for mass adoption.

In conclusion, while bio-inspired designs hold significant promise for improving the efficiency, sustainability, and performance of automobiles, the path to widespread adoption is fraught with challenges. Addressing issues related to material availability, technological compatibility, cost, and regulatory compliance will be critical in

Table 10.3 Benefits, challenges, and future prospects.

Sl. no.	Domain	Details
1	Advantages of automobiles	Drag is reduced; therefore, fuel consumption is less. The emission is low and eco-friendly. There is more stability and safety.
2	Environmental impact	Minimum energy/optimized energy vehicles. Sustainable technologies inspired by natural organisms, animal gait, and aerodynamics.
3	Cost benefits	Fuel cost is saved; durability and lifespan are improved.
4	Challenges	Biological processes are complex and difficult to fabricate in design.
5	Future prospects	Real-time adaptation with AI integration, especially in electric vehicles, will extend to other industries of rail, marine, SUVs, trucks, and fluid machinery.

overcoming the barriers to innovation in the automotive industry. Table 10.3 shows the benefits, challenges, and future prospects.

10.4 Conclusions

Nature-inspired automotive design is a growing trend that seeks to harmonize the demands of modern engineering with the sustainable, efficient principles found in the natural world. Through the application of biomimicry, automakers are developing vehicles that are more efficient, environmentally friendly, and innovative. From lightweight materials to aerodynamic designs and regenerative energy systems, nature's solutions provide a roadmap for the automotive industry's future. As technological advancements continue, nature-inspired designs will likely become a cornerstone of the automotive industry, helping to meet the challenges of sustainability and performance in the 21st century.

This chapter classifies research on biomimetics into five sections. The classifications are aerodynamic, structural, wheel assembly, power train applications, and applications in digital manufacturing. There are two clusters: microscale and macroscale applications. Macroscale biomimetic structures have so far infiltrated the powertrain, wheel assembly, undercarriage, and car body. The main goals of microscale research were to improve wear resistance, lubrication, and surface qualities. These advancements include enhancing the vehicle's performance, boosting fuel efficiency, and adding new characteristics to next-generation cars like wear resistance, self-cleaning, and self-healing. The microstructural implications have been more involved in recent studies. Their capacity for autonomous implementation is the primary cause of this. In other words, even after they are completely produced, microstructural materials can be developed as coatings and used in automotive advancements. However, the macro-scale advances show the biggest impact of biomimetic design techniques. This paper will serve as motivation for future scholars to become more active in macro-scale developments. To sum up, biomimicry is a promising technology that is becoming more popular for upcoming car advancements. To comprehend current trends and get fresh ideas for further research, the authors think that this review paper will be helpful to researchers working in the automotive industry.

10.5 Future Research Opportunities

Based on recent research, biomimetic designs have improved several car parts by adding futuristic elements. To put it briefly, biomimicry has improved the micro- and macro-structural aspects of cars. Wheel assembly components, powertrain and under-the-hood components, and vehicle body components are all included. These enhancements have improved the car's performance while using less petrol. However, there are still areas in car design that require more focus, but were only identified in earlier studies. This covers the under-hood parts, wheel assembly parts, and interior parts. As a result, the researchers observed a significant study gap in the chosen area. Therefore, biomimetic solutions for automotive issues have yet to be found.

Vehicles have widely incorporated a variety of complex biomimicry-based shapes into their external designs. However, more traditional Cartesian geometries, like planes, cylinders, and spheres, are typically used in the design of mechanical parts found beneath the hood, such as fans, ducts, and heat exchangers. As a result, these designs are heavy and require a lot of engine bay packing area. On the other hand, biomimicry has only influenced the parts of the car's interior that engage with humans. The interior of the car is comfortable and easy to use by the rider, even though the exterior and power train are the car parts that have the greatest effect on car performance. Therefore, when buying a car, the inside environment is a crucial factor to consider. The concern regarding bio-inspired solutions for the more substantial automobile interior parts is anticipated to increase in the future.

The anti-wear mechanism found in mollusk shells (conches) can be applied to automobiles in several ways that were not apparent in the designs related to automobiles. This mechanism is developed in seashells to act against water-sand erosion on sandy beaches and is a result of a combination of their unique shape and bio-tissue. New research shows that molluscan shells are bio-ceramic composites that, over millions of years, have developed very complex microstructures because of evolving biotic interactions within the environment. Another marvel of biotechnology is mimicking chameleon skin properties. Notably, chameleon skins have color-changing capabilities, and this technology can even be applied to automobiles with truly futuristic forms. Studies have already shown evidence of altering the microstructural characteristics of magnesium-based alloys to impart different colors to the surface while maintaining a reasonably high mechanical performance. Nonetheless, it appears that nature-inspired car designs have a bright future. The potential for further incorporating biomimicry into vehicle design is enormous, given the ongoing developments in AI, renewable energy technology, and materials research. We anticipate that nature-inspired solutions will become increasingly important in the automobile sector as the demand for environmentally friendly and energy-efficient mobility increases globally.

Bibliography

1. Yin., H., Zheng., X., Wen., G., Zhang., C., Wu., Z., Design optimization of a novel bio-inspired 3D porous structure for crashworthiness. *Compos. Struct.*, 255, 112897, 2021.
2. Nikkhah., H., Baroutaji., A., Kazancı., Z., Arjunan., A., Evaluation of crushing and energy absorption characteristics of bio-inspired nested structures. *Thin Walled Struct.*, 106615, 2020.
3. Zhang., W., Xu., J., Yu., T.X., Dynamic behaviors of bio-inspired structures: Design, mechanisms, and models. *Eng. Struct.*, 265, 114490, 2022.
4. Zhang., W. and Xu., J., Advanced lightweight materials for Automobiles: A review. *Mater. Des.*, 221, 110994, 2022.

5. Shaharuzaman., M.A., Sapuan., S.M., Mansor., M.R., Zuhri., M.Y.M., Conceptual Design of New Bio-inspired Automotive Side-Door Impact Beam. *Int. J. Eng. Adv. Technol.*, 9, 3, 2020.
6. Ramasubramanian., S. and Tennis., K., Lightweight material for weight reductions in an automotive suspension part lower link. *Mater. Today Proc.*, 2023.
7. Wang, S., Yu., X., Shen, L., Yang, A., Chen, E., Fieldhouse, J., Barton, D., Kosarieh, S., Noise reduction of automobile cooling fan based on bio-inspired design. *J. Automob. Eng.*, 235, 2-3, 465–478, 2021.
8. Kim., D., Lee., H., Yi., W., Choi, H., A bio-inspired device for drag reduction on a three-dimensional model vehicle. *Bioinspir. Biomim.*, 11, 026004, 2016.
9. Jae Kim., J., Hong., J., Joon Lee., S., Bio-inspired cab-roof fairing of heavy vehicles for enhancing drag reduction and driving stability. *Int. J. Mech. Sci.*, 131–132, 868–879, 2017.
10. Toong Foo., C., Oma., B., Taib., I., Shape Optimization of High-Speed Rail by Biomimetic. *MATEC Web Conf.*, vol. 135, p. 00019, 2017.
11. Mudassir, M., Tarlochan, F., Mansour, M.A., Nature-Inspired Cellular Structure Design for Electric Vehicle Battery Compartment: Application to Crashworthiness. *Appl. Sci.*, 10, 4532, 2020.
12. Igor., G., Viera., K., Štefan., M., Application of biomimetic design in concept of small terrain electric vehicle, MODULO. *J. Eng. Res.*, 2021.
13. Gm., A., Application of Biomimetics in Design of Vehicles – A Review. *Int. Res. J. Eng. Technol.*, 07, 3, 2753–2759, 2020.
14. Wijegunawardana, I.D. and de Mel, W., Biomimetic Designs for Automobile Engineering: A Review. *Int. J. Automot. Mech. Eng.*, 18, 3, 9029–9041, 2021.
15. Aili., Z. and Long., J., Research on the Bionics Design of Automobile Styling Based on the Form Gene. *IOP Conf. Ser.: Mater. Sci. Eng.*, 234, 012020, 2017.
16. Liu, C., Wang, G., Zhou, H., Mei, Y., Li, Y., Li, Y., Zhang, L., Design of Two-Wheeled Motorcycle Tire Crown Contour Bioinspired by Cat Paw Pads. *Appl. Bionics Biomech.*, 2020, Article ID 8898413.
17. Loganathan., B., Ahmed., T., Mustary., I., Mahmud Sonyy., S., Alam., F., Aerodynamic design of an energy-efficient light truck by biomimicry. *AIP Conf. Proc.*, vol. 2681, p. 020101, 2022.

18. Patel, A. and Braae, M., Rapid Turning at High Speed, Inspiration from the Cheetah's Tails. *IEEE/RSJ International Conference on Intelligent Robots and Systems*, 2013.
19. Pleasant, D., Gavin, C., Redden, G., Nagel, J., Zhang, H., Bioinspired Design of Material Architecture for Additive Manufacturing. *Machines*, 11, 12, 1081, 2023.

Index

β-sheet crystals, 188

3D printing, 120

ACO, 57
Actuators, 180
Adaptive control, 225
AddACO, 62
Additive, 57
Additive manufacturing, 13, 38, 175
Aerodynamics, 2
Aerospace structures, 52
Air monitoring, 221
Amorphous domains, 188
Anisotropic adhesion, 148
Ant system, 60
Ant walk, 70
Anti-biofouling, 167
Anti-icing, 102
Applications of Gecko, 140–141
Artificial spider, 195

Battery capacity, 215
Beetle, 86
Bioactive, 254
Bioactivity, 115
Bio-assisted manufacturing, 33
Bioglass, 118
Bio-inspired, 57, 85
Bio-inspired design, 1, 31, 210
Biomimetic adhesives, 136
Biomimetic implant, 111
Biomimetics, 6, 33
Biomimicry, 241
Biomineralization, 47
Bionics, 6
Bioprinting, 44
Body shapes, 254
Bonerepair, 110
Bristling effect, 169
Brittleness, 115

Cactus, 96, 254
CAD modeling, 174
Capillary forces, 137–138
Carbon fiber, 249
Cassie-Baxter model, 95
CFD analysis, 169
Climbing fobots, 154–155
Collision avoidance, 227
Composite materials, 37, 119
Contact angle, 99
Control systems, 224
Cooling system, 258
Corrosion resistance, 115
CPP, 73
Crystalline-amorphous, 188

Decentralized, 65
Degradable implants, 116
Denticles, 167
Digital modeling, 174
Directional adhesion, 148–150
Dolphin, 243
Dragline silk, 189
Drag reduction, 164, 241

Drones operation, 210
Dry adhesion, 136
Dynamic wetting, 96

Electric vehicles, 248
Electrospinning, 16, 31, 89, 112, 140, 177
Energy efficiency, 247
Environmental monitoring, 221
Evolutionary adaptations, 144–145
Exoskeleton, 87

Fabrication methods, 122, 170
Fabrication techniques, 140
Finite element, 113, 174
Flapping wings, 218
Flexible substrates, 178
Flight agility, 210
Flight efficiency, 229
Flight mechanics, 214
Flight stability, 211
Flow control, 243
Flow sensors, 179
Fog harvesting, 24, 102
Frictional adhesion, 145–146
Fuel saving, 254
Functionally graded materials, 40

Gecko feet, 2
Gecko toe pads, 142
Glycine-rich domains, 188

Hierarchical structure, 138–139
HMM, 70
Honeycomb, 249
Hovering flight, 214
Hydrodynamics, 168
Hydrogels, 14, 49
Hydrogen bonding, 188
Hydrophilic, 86
Hydrophobic, 86, 250
Hydrophobicity, 146
Hydroxyapatite, 7
Hydroxyapatite coating, 118

Industrial scalability, 205
Inkjet printing, 90
IoD, 73

Laser etching, 91
Laser techniques, 88
Leaf, 254
Lift enhancement, 165
Link predication, 72–73
Lithography, 41
Load distribution, 113
Localization tracking, 216
Lotus effect, 250
Lotus leaf, 2

Major ampullate, 189
Maneuverability control, 211
Manufacturing techniques, 170, 218
MaSp-2 protein, 190
Material selection, 114, 152, 218
Materials, 93, 248
Mathematical modeling, 97, 124
Mechanical properties, 168
Mechanical strength, 114
Micro UAVs, 210
Microfabrication, 140
Microfluidics, 45
Microneedles, 21
Microstructures, 87
Microstructuring, 177
Miniaturization challenge, 210
MMAS, 61
Monte Carlo, 67
Morphing wings, 18
Motor weight, 218
Multi-material printing, 40
Mutualism, 63

Nacre, 7, 51, 250
Nanocoating, 119, 179, 254
Nanofibrillar structure, 191
Nanoscale spatulae, 139–140
Nanostructures, 87
Noise reduction, 249

Optical transmission, 201
Optimization, 100, 254
Optimization algorithm, 229
Orb-weaver spiders, 188
Origami-inspired design, 48
Osteointegration, 114

Particle image, 169
Path planning, 59, 215
Perching mechanism, 230
Pheromone, 71
Photo reconfiguration, 170
Photolithography, 41, 92, 140, 177
Pitcher plant, 96
Poly-alanine blocks, 188
Polydimethylsiloxane (PDMS), 140
Polymer, 93
Polymer composites, 165
Polymer implants, 118
Porosity optimization, 129
Primary structure, 187

Quantum-based optimization algorithm, 72

Regulations policy, 220
Reversible adhesion, 148–149
Riblets, 165
Robotic grippers, 154

Scaffold design, 121
Scalability, 101
Scansors, 138
Scheduling, 57
Secondary structure, 187
Self-assembly, 42
Self-cleaning, 250
Self-cleaning adhesion, 135
Self-cleaning surfaces, 14
Self-healing, 8
Sensor calibration, 215
Sensor systems, 215
Sensors, 179
Setae, 142–143
Shape memory alloys, 2, 181
Shark skin, 15, 243
Skin–core structure, 192
Smart materials, 2, 120
Smart sensors, 179
Soft robotic systems, 153–154
Sorting, 67
Spatulae, 142–143
Spider silk, 15, 51, 96, 185
Spidroin, 187
Stimulus, 77
Stimulus-responsive materials, 14
Strength synergy, 186
Stress analysis, 123
Stress shielding, 110
Stress–strain characteristics, 167
Supercontraction, 190
Superelasticity, 186
Surface coatings, 178
Surface energy, 99, 146
Surface textures, 243
Surveillance reconnaissance, 220
Sustainability, 1
Swarm, 57
Swarm algorithm, 229

Task allocation, 77–78
Termite mounds, 18
Thermal comfort, 254
Thermal conductance, 201
Threshold, 77
Tire grip, 255

Tissue engineering, 44
Titanium alloy, 112, 117
Torsional damping, 202
Toughness, 186
TSP, 58

Van der Waals adhesion, 8
Van der Waals forces, 135
Velcro, 6
Vortex control, 254

Water scarcity, 85
Wearable electronics, 140–141
Weight reduction, 249
Wenzel model, 95
Wettability, 94

Printed and bound by CPI Group (UK) Ltd, Croydon, CR0 4YY
17/09/2026
14964283-0001